Smart Structures and Materials

For a complete listing of the *Artech House Optoelectronics Library*,
turn to the back of this book.

Smart Structures and Materials

Brian Culshaw

Artech House
Boston • London

Library of Congress Cataloging-in-Publication Data
Culshaw, B.
Smart structures and materials/Brian Culshaw
p. cm.
Includes bibliographical references and index.
ISBN 0-89006-681-7
1. Smart materials. 2. Smart structures I. Title
TA418.9.S62C85 1996 95-48914
620–dc20 CIP

A catalogue record for this book is available from the British Library.

The cover photo, courtesy of the Kajima Corporation, depicts the KaTRI No. 21 building, a variable-stiffness, earthquake-resistant building in Tokyo.

685 Canton Street
Norwood, MA 02062

International Standard Book Number: 0-89006-681-7
Library of Congress Catalog Card Number: 95-48914

10 9 8 7 6 5 4 3 2 1

Contents

Preface

The concept of a "smart" structure or material is a total anathema to some and a technical and intellectual stimulus to the majority. The concept is all about materials and structures that can react to the world within which they operate and thereby enhance their functionality or survivability. The concept is wide ranging in both its applicability and its technological basis and principles. The applications arena encompasses virtually everything that is engineered—everything where there is a structural material or structural function. So smart structures and materials will find their way into aircraft, trains, civil engineering structures, domestic appliances, and, in time, toys and the amusement arcade. The necessary technological armory is also wide ranging: the smart structure needs inputs from control engineers, signal processors, sensor and transducer engineers, actuator designers, materials scientists, and structural designers. The realization of the smart material requires this and more—materials design and processing and molecular-level engineering.

The subject is then truly interdisciplinary, while this book presents one person's view. Any book has to be written in a finite time and to a finite length, so my aims in writing this book have been to introduce what I consider to be the essential disciplines, to endeavor to bring these disciplines together, to illustrate the integration of these concepts into structural assemblies, and to stimulate the reader to find more applications for these ideas and to burrow into complementary texts and other sources of information to complete his or her required picture of the subject. The intent is to give a feel for what is needed, to give a feel for what is available at the present time, and to speculate on what this may all imply for the future.

Structures that respond can call for the repair person; react against wind, wave and earthquake; adjust their thermal properties; breathe and cool themselves; compensate for disturbances or vibrations; adjust themselves for wear; issue warnings; and ring alarms. Eventually, they will design themselves and take the engineer's responsibilities from the technical to the aesthetic and optimize the integration of the structural, the functional, and the intelligent into a synergistic entity.

The subject of smart structures and materials is all about engendering engineered "things" with intelligence—or at least with the ability to react to circumstances to enable a more cost effective survival. The engineered "thing" can be anything from a bridge or an airplane to a plate (to warn if your meal is really appropriate) or a washing machine.

Of course you could argue that many of these are somewhat intelligent already, and indeed this is the case. So why the fuss? My own view is that the availability of ever-expanding manmade intelligence—not only in computers but also in sensing and measurement systems and in generalised actuation functions—will permeate the way in which the current traditional engineer must think about design and realization. Within a very short time, this will modify his or her total philosophy and swing professional responsibilities more and more from the technical into the social. The technologies that permit this to happen are already emerging.

In the book I have scratched the surface of these contributory technologies, including sensing, actuating, structural engineering, mechanics, control, and signal processing. The coverage is inevitably patchy, but I hope the control engineer will not be too frustrated by seeing his or her own very important domain given a very short treatment. I also hope that the fiber-optic sensor specialist doesn't interpret my own (admittedly prejudiced) account of the subject as an authoritative statement that fiber optics is the only way to instrument a structure. It isn't. The subject is totally interdisciplinary, so inevitably one person's account is presented from his or her own perspective. I shall leave the reader to draw his or her own conclusions.

When I considered the subject of smart materials, I wondered whether to include anything at all on this curious and ill-named topic. The implication within the term "smart material" is that something synthetic posses the ability to respond to its environment using the deductive rather than instinctive reactions. My own view is that such materials as straightforward, pure compounds cannot exist—although when fabricated as mixtures and hybrids they could perhaps aspire to mimic biological models. The aim of the short chapter on the subject is to explain why—and also what—is needed in order to aspire to these dizzy heights. I also question why we need smart materials, since surely engineers and scientists should aim to complement human skills rather than emulate them. Historically, this is exactly what we have always done, though of course the future may be different.

The book contains some brief case studies that indicate some of the very many directions in which the topic may go and towards which it may be applied. These case studies also attempt to indicate—without entering the dangerous business of quantifying in dollars, yen, or pounds—the benefits of smart structures technology in terms of enhancing reliability and improving performance. The book omits a lot—it was edited by time, pages, and patience. At the technological level, I have opted toward an attempt to give a feel for the subject rather than enter gory detail. At the applications level, the examples are a very few of a very many. I have missed out much of the current thinking in aerospace, of adaptive shaping of lifting and control services, of variable radar signatures, of distributed electronics. I have omitted the Glasgow School of Art's imaginative project on smart materials, which speculated on self-resurfacing roads, adaptive glazing and insulation, and the smart refuse truck. I really haven't talked about polymer electronics, and I even omitted mention of our own work on construction process control monitoring. And smart prostheses are already being tested in sheep: these carbon fiber and piezoelectric hybrids are an exciting amalgam of biological principles, materials science, and electronic materials.

What remains is, I hope, enough to convey the flavor and stimulate the imagination. The latter is particularly important. The ideas that are outlined in this book will influence everything from bioengineering through transport technology, through structural engineering to architecture. In time the concepts will influence our abilities in medical implants and promise to realize a more cost-effective and socially responsible society. Far reaching claims! Since smart structures concepts will pervade almost every nook and cranny of engineering and technology—and I believe the process is inevitable—all we can dispute is the timescale. We are seeing the burgeoning information technology industry beginning to be applied to good effect in other sectors. The essence of the interest in smart structures and materials is that these other sectors, which have progressed but slowly, are now appreciating that technological advances coupled with the available information technology can realize very substantial benefits and on relatively short timescales.

The book has benefited from an awful lot of help from others. The anonymous reviewer contributed enormously with cryptic comments, helpful suggestions, and broad hints and, in particular, helped to rationalize some of the underlying logic. I should also thank friends and colleagues within my own research group and within the many collaborative projects that we have undertaken in the smart structures area. Peter Gardiner contributed greatly with his insight into application potential and through countless enthusiastic discussions. The many participants—both lecturers and attendees—on our smart structures and materials courses have provided a healthy backdrop and some very comprehensive background information. Aileen Mitchell turned dictation into a respectable manuscript with skill, patience, and a not uncritical eye. The incessant, and constructive, reminders from Julie Lancashire and Kate Hawes at Artech ensured the manuscript was actually delivered, rather than simply promised. My family tolerated once again the eccentricities and self-imposed solitary confinement of the author.

I hope that this book—warts and all—will convey at least some of the technical excitement and real potential of an important emerging discipline. I learnt a very great deal through writing it and became all the more optimistic for the future by the time I had finished. There is still much to be done but the seeds are sown on fertile ground. Exactly what will grow remains something of a mystery. We shall see.

Brian Culshaw
Glasgow
December 1995

Chapter 1

Smart Structures and Materials: Why Are They Interesting and What Are They?

This book is about smart structures and materials. In it we shall investigate why smart structures are interesting, what is meant by smart structures and materials, how they may be conceived and turned into a reality, and what they offer to present and future engineering technologies.

1.1 WHY ALL THE EXCITEMENT?

Those of us who have sat in traffic lines approaching major roadworks or who have waited patiently while some vital part of an aircraft is restored to a safe operating condition have inevitably questioned whether the engineering profession can improve the way in which it monitors, controls, and adapts structures to changes in local environmental and loading conditions. We all admire the ability of the chameleon to blend into the background. But we also aspire to similar adaptability in synthetic materials—imagine the self-coloring carpet optimizing its design to its immediate environment. This is entirely where smart structures and materials come into play. For example, in large civil engineering structures such as bridges (Figure 1.1), an integrated instrumentation system could tell both the owner and the users of the structure an enormous amount about both the environmental and traffic-loading conditions and the general condition of the bridge—in particular, when and whether it was in need of repair. An entirely active building could take the whole idea a stage further so that, for example, in high winds or under earth tremor conditions, the building could adjust its dynamic properties in order to minimize both discomfort to the occupants and the possibility of structural damage. Such smart systems are beginning to be realized, and indeed the concept of structural measurement for performance monitoring has been with us for decades. However, a radical change has occurred in the past few years due to immensely improved available

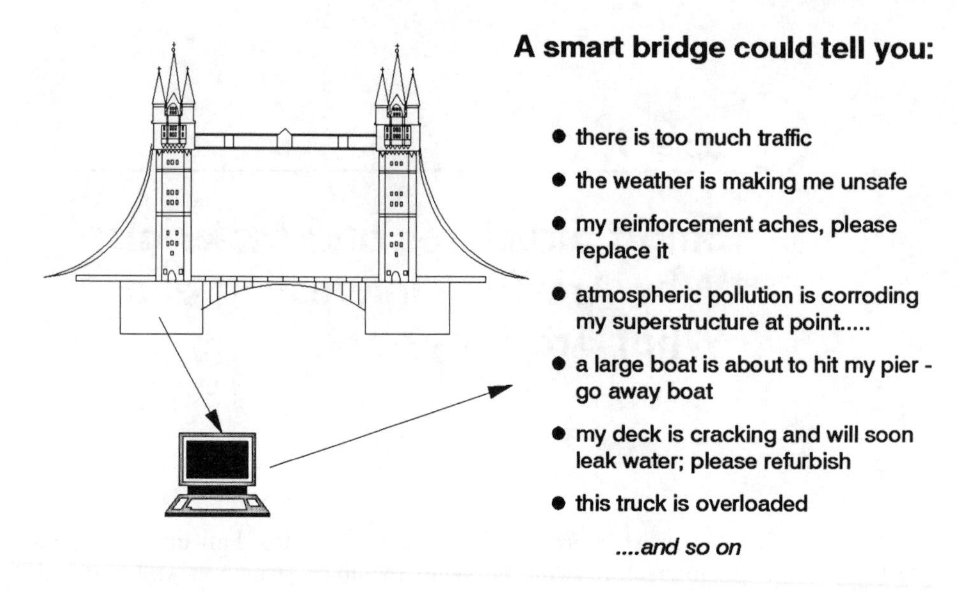

Figure 1.1 Some of the possible gains from using smart concepts approaches in major civil engineering structures.

instrumentation (most notably computing power, new transducer and measurement concepts, and interference-free optical fiber transmission) that offers the potential of achieving in the field what was previously only possible in laboratory models.

There is even greater scope for active self-testing instrumentation, which in effect incorporates self-diagnostic technology into the structural fabric of a machine to monitor the onset of fatigue cracks and structural damage, corrosion, and erosion. Possibly the most attractive single applications area for such ideas (Figure 1.2) is in transport systems. The ever-increasing use of new materials, becoming both lighter and stronger (most notably glass and carbon fiber reinforced plastics), is accompanied by nagging doubts in system designers and users concerning the failure mechanisms of these complex material structures. Further, all transportation vehicles are subject to very conservative servicing and maintenance schedules so that an intelligent vehicle that was only repaired when it needed repairing could be economically very attractive.

These examples rely upon applying advanced instrumentation and signal processing and recognition systems to relatively conventional structures. Even in this area there is a multitude of possible applications now that the cost-performance benefits ratio are beginning to be recognized and increasingly stringent legislation (especially in the environmental and pollution areas) imposes the need for increasingly sophisticated monitoring systems. The desirability for smart structures and all this implies has been superseded by a real need.

A fully instrumented "smart" plane, boat, car, or train could tell you:

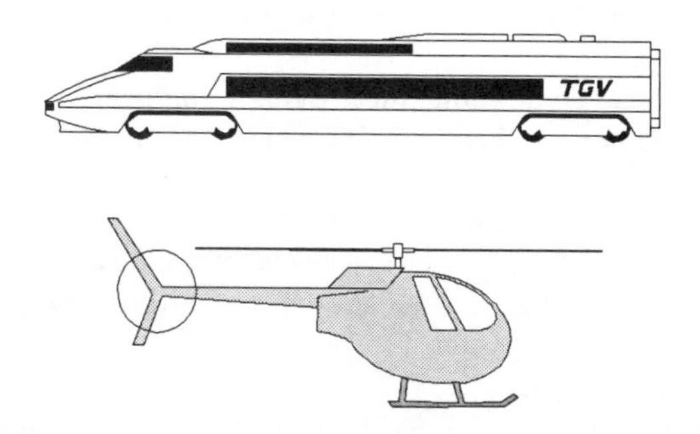

- **I don't need servicing yet**
- **my rhs upper body panels are corroding at point...**
- **someone hit me with a stone at point...**
- **my fuel tank will soon be leaking**

and so on... or even

- **all is well, I'm feeling fine!**

Figure 1.2 The active smart structure (self-testing and diagnosis) has great potential in transport and safety critical systems.

As for smart materials, the need for and the status of this particular discipline is open to much more debate. Many regard straightforward transducer materials as "smart," but as we shall see in the following section, such materials really fail to meet any criteria for "intelligence"—although this is not to dismiss in the slightest their very considerable importance. My view is that intelligence implies an adaptive response that is conditioned to acknowledge a number of input stimuli. Such material *systems* do exist and invariably involve a spectrum of constituents. The diversity of material structures used in engineering systems is immense, embracing everything from the piezoelectric pressure transducer to the shape memory alloy actuator to the photochromic window and the tunable electromagnetic absorber.

The focus in this book is primarily on structures and structural instrumentation—the domain of the electronic, mechanical, civil, and structural engineering communities. This does, of course, encompass instrumentation systems and advances in materials for sensing and actuation. However, we shall limit the discussion of intelligent material systems to a very brief explanation of their potential and of the challenges that such systems prevent. Before going into this, it is important to establish the linguistic framework that surrounds this relatively new and somewhat speculative discipline.

1.2 WHAT THE DICTIONARIES SAY

The terms "smart structures" and "smart materials" are much used and more abused. The starting point for this book must then be to investigate the meanings of the terms and to attempt to establish a framework of actual and potential realities within which smart structures and smart materials may be conceptualized. Of course, it is always a gamble to establish boundary conditions because, sooner or later, someone, somewhere will venture outside; however, it is my total conviction that the discussion must start by establishing such a framework.

The English language must provide some guidelines, though engineers often forget the dictionary and evolve a language of their own. Here is what the Shorter Oxford says:

- *Smart:* severe enough to cause pain, sharp, vigorous, lively, brisk . . . clever ingenious, showing quick wit or ingenuity . . . selfishly clever to the verge of dishonesty;
- *Material:* matter from which a thing is made;
- *Structure:* material configured to do mechanical work . . . a thing constructed, complex whole.

For once, Webster's is not too dissimilar, though surprisingly *the IEEE Dictionary of Electronics* in its 1984 edition makes no attempt to define any of these terms. But does the English language help?

1.3 SMARTNESS—THE ENGINEER'S APPROACH

In the engineering sense, smart systems make sense of complex, often surplus, information. Perhaps Shannon would argue that surplus information is simply inefficiently coded information, and this point, while it may encroach within the realms of philosophy, contains an important hidden message that would perhaps encompass the role of the smart system in the simple statement that it imposes a reduction in local entropy. Quite how this relates to the Shorter Oxford is a topic for debate and, while it is tempting to use the last dictionary definition, probably the middle one is the most appropriate. Interestingly no treatise on information theory that I have encountered has extrapolated "smartness" into this somewhat abstract definition, though I believe that viewing "smartness" in entropy terms can shed some important light upon the subject.

The first important point is that a local entropy reduction fundamentally requires some input of energy from an outside source, so all smart systems have some form of energy source associated with them. When we discuss smart materials, this is a basic, fundamental feature.

Information reduction is, however, the major issue. This necessity for information reduction eliminates a whole host of functions from the definition "smart." For example, the whole area of measurement systems is one that recurs in smart technology. A simple pressure transducer that produces a voltage dependent upon the input pressure in a direct one-to-one relationship could never be regarded as "smart." However, a pressure transmitter incorporating a thermocouple that measures both temperature and apparent pressure and corrects the apparent pressure taking due note of the sensor's temperature coefficient could be regarded as smart. It takes in surplus information (the temperature) and reduces it to a single value of corrected pressure. How this process is actually implemented is irrelevant. Many pressure transducers apply correction factors within the structural design by the ingenious use of differential thermal expansion coefficients within materials. Here the material structure effectively performs information reduction by measuring the temperature and turning it into a mechanical force that corrects the apparent pressure. In this case the source of energy that is used to perform the corrections is the actual temperature fluctuation itself—we have a thermally driven computer.

While we could content ourselves with this relatively tight and limited definition of "smartness," the engineering community usually wishes to go further and asks the question, "How should we use this information when we have it?" Consequently, part of a smart system invariably implies some form of feedback with decisions made on the basis of information received and used to correct the performance of the system under consideration. At the conceptual level, we then have a possible representation of the smart system as shown in Figure 1.3 where multiple channels of information are examined and assessed to provide a signal that finally modifies a response process.

1.4 STRUCTURES

Here I believe the dictionary is a genuine help in referring directly to the ability "to do mechanical work," though perhaps the dictionary is a little unscientific since the ability to perform mechanical work implies the ability to convert an input energy source into the mechanical output. This, however, is not so constricting as it may

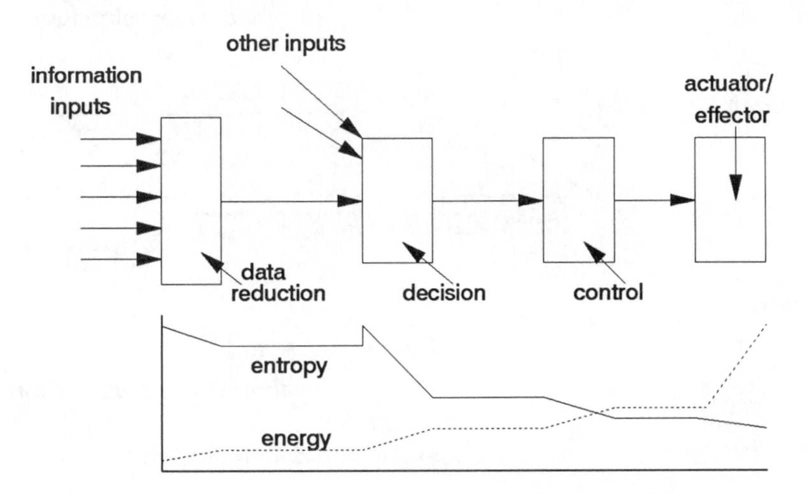

Figure 1.3 Information flow in a smart structure or material.

first appear since, for example, a bridge will store the potential energy introduced by the very slight depression imposed by a passing truck and expend that potential energy in restoring its shape to the unloaded form when the truck has disappeared. Therefore, anything that is capable of bearing a load may be termed a structure. Perhaps, though, it is more useful to regard anything that is *designed* to take a mechanical load as a structure. Within this definition we have everything ranging from overhead power cables to bridges to aircraft to mechanical gearboxes to highways.

A *smart structure* is therefore one (Figure 1.4) that monitors itself and/or its environment in order to respond to changes in its condition. For example, it may be self-repairing (or indeed request some external agency to do the repairing for it), or it may use variable stiffness elements to control its response to applied mechanical loads. The smart structure as sketched in Figure 1.4 is by no means a new idea. Indeed, most of the concepts that are discussed in this book are well established. All that has changed is the technological capability to turn a concept into a reality. Aircraft are a prime example. They have, since the very early days, responded to loading conditions by using wing flaps and the like to correct for the air-flow characteristics at low speeds. Recent and emerging technological advances have immeasurably broadened the scope of the smart structure, offering exciting possibilities in everything from civil engineering to nanotechnology.

1.5 MATERIALS

The dictionary definition is perhaps less helpful since it requires that "a thing be made," though I find it difficult to conceive of any material from which a thing is

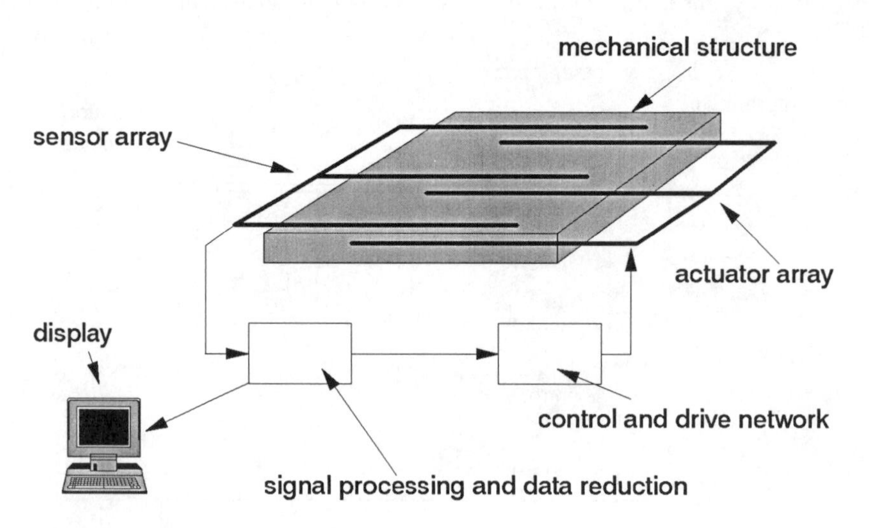

Figure 1.4 Conceptual diagram of the "smart" structure.

not made—even if it is nothing more than compost. That said, there is an implication that a material is a solid despite the obvious point that "things" can also be made from liquids and gases by combining them together in an appropriate chemical formulation. Conventional wisdom does, however, limit materials to solids. Technologists rarely deal with liquids or gases unless it is to deposit from the fluid state into a solidified form. Classifying materials is always dangerous, but it is tempting to distinguish "structural" and "other" materials. This is a useful distinction but, as we shall see shortly, a somewhat dangerous one.

In the "other" materials category lies a whole host of possible functional materials. These may be chemically reactive, they may have controlled optical or radio frequency properties, or perhaps they are conductive or semiconductors or ferromagnetic. The range of nonstructural materials is enormous.

All materials are also responsive. Whether or not they are *smart materials* is a different question. This responsivity is often useful and forms the basis of transducer technology. Figure 1.5 illustrates the functions of a range of transducer materials and indicates how their input/output characteristics may be summarized in a simplistic graphical form. However, returning to our original definition, there is no "smartness" here since the output has the same, or sometimes a larger, information content as the input. To engender smartness we need responsivity to a second variable, illustrated conceptually in Figure 1.6. If a material can be *designed* to produce a specific response to a combination of inputs, then it will fulfill the "smartness" requirements. Indeed, the traditional pressure transducer designs exemplified in the early part of this chapter can certainly be construed as a smart material structure. However, it begs the question of whether a *single* material may ever be construed as smart since, at least to date, all mechanically responsive analogue computer systems—like the pressure transducer—are of hybrid design.

Another observation that is often useful is that "other" materials are invariably mechanically soft. Historically, the engineer has built his structures from stiff, hard materials and, indeed, has been very successful in doing so. The "softer" end of materials have tended to be more overtly responsive to environmental influences but have usually had poor structural properties. Perhaps this responsivity can be built into a traditional "hard" structure to the benefit of both. This, the integration and exploitation of hard and soft materials, is one of the emerging challenges in the study of smart structures and materials that as yet has been barely addressed.

1.6 HYBRID STRUCTURES—THE SOFT AND THE HARD

It rapidly becomes evident that anything that is "smart" and "structural" will require a combination of hard load-bearing materials and soft invariably mechanically irrelevant information-processing materials. This is hardly surprising since the much vaunted concept that smart structures must mimic biological precursors follows

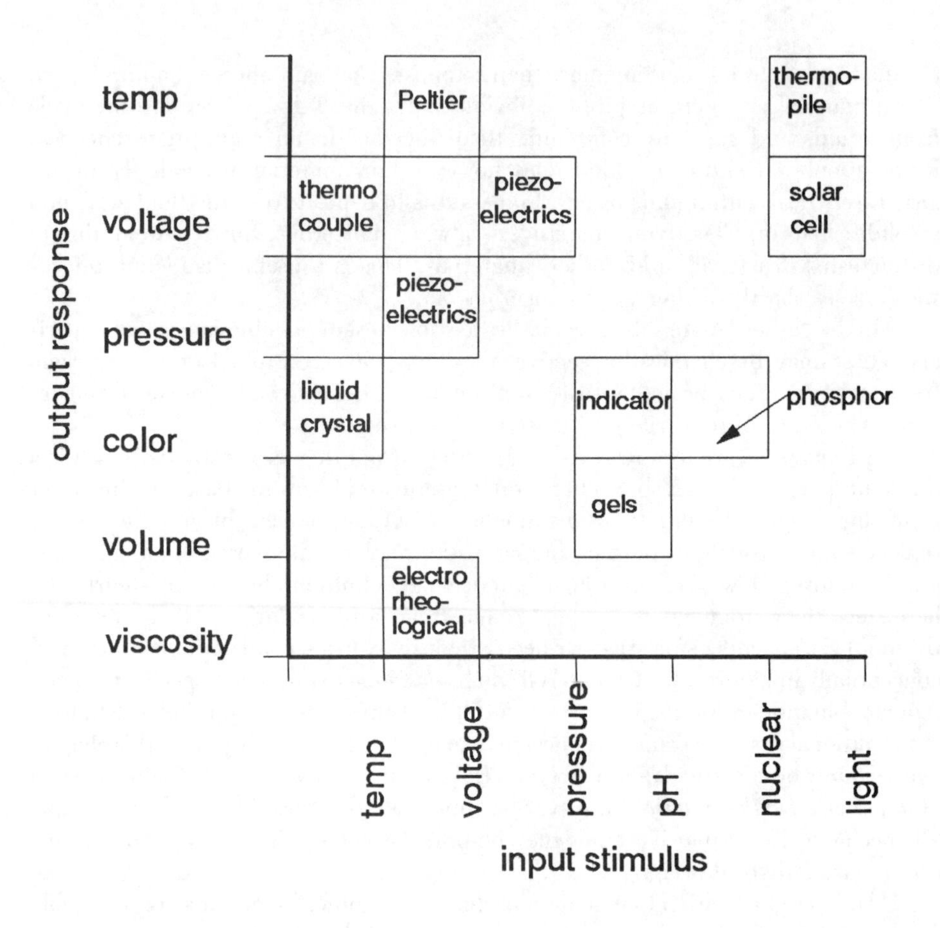

Figure 1.5 Responses of some sensor materials. These materials are often referred to as "smart," but are they?

exactly along these lines. However, caution is essential. In evolution, large natural structures have tended to be inherently dumb. Probably the largest natural structures that are not entirely passive are the giant redwoods of California—though their intelligence is limited to reacting to the environment, to producing redwood cones, to seeking the sun, and sometimes to self-healing and growth. There are other severe limitations as well as size upon natural biological precursors. In particular, the smart creatures of the animal kingdom all survive in relatively carefully controlled thermal environments and find it very difficult to cope with abnormal physical loads. The beached whale, for example, suffocates under its own weight.

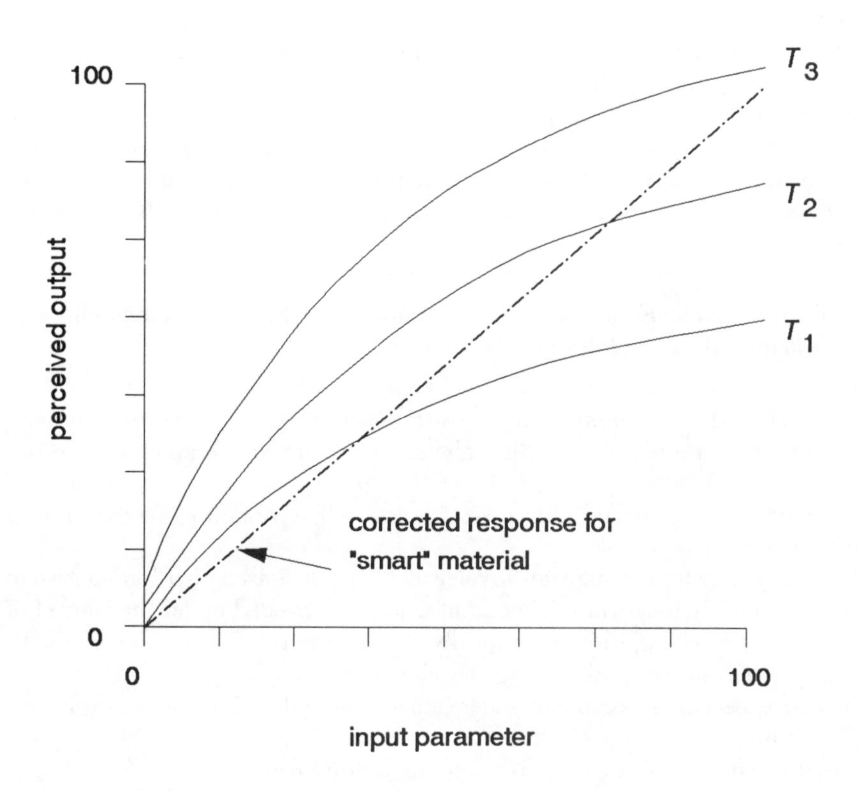

Figure 1.6 Illustrating the response of a "smart" material. The uncorrected response is a function of the parameter *T*, but after correction in the material, an unambiguous output is obtained, independent of *T*.

This is not to decry the value of the animal kingdom and the significance of what can be learned from studying biological systems. While the features of biological systems are many and varied, I believe that the principal points that may be gleaned from biological precursors are as follows.

Biological structures that are dominated by wood and bone demonstrate the following characteristics:

- The useful strength-to-weight ratio of a structure may be enhanced by making it cellular and usually essentially fibrous. This approach gives very high strengths

in specific directions, often in longitudinal compression, at the expense of fragility to unusual forces.
- Self-repairing systems may be realized but do require local information processing and a supply of usable energy.
- Adaptive jointing mechanisms designed to hold the entity together can be important (mainly muscles and tendons) in permitting the movement of the overall structure and in allowing the structure to sustain unusual loads; that is, skeletons may be viewed as variable stiffness structures with stiffness control through muscular action.

Intelligence in biological systems is little understood, but biosystem intelligence can be characterized by the following.

- The processing is distributed with a central overseeing and monitoring system.
- The communication channels throughout the system are complex and reconfigurable.
- The central processor is highly adaptive and self-reconfigures in the light of experience.
- The energy transfer mechanisms involve a chemically based distribution system that is selectively burned as, when, and where it is needed under the control of the central processing unit using locally derived information involving stimuli from external sources both locally and elsewhere.
- The entire system is capable of self-regeneration and maintenance and is also self-cleaning.
- Its performance as a processor is very temperature dependent.

In an engineering sense, the lessons to be learned from biological precursors may be simply stated:

- Use a "hard" structural configuration and tailor its mechanical symmetry either mechanically during its formation processes or through some quasi-biological growth mechanisms in order to optimize its strength to the anticipated load distribution.
- Incorporate adaptive load-bearing capacity, that is, muscles, into the design of the structure to enable it to move as required and to cope with unusual loading conditions.
- Optimize the energy transfer and control processes and the energy source for the muscular system and for the local information processor.
- Ensure that the intelligence that controls the structure can respond to both internally generated and externally imposed stimuli and can adapt responsively to these in appropriate combinations.

Engineering has evolved in its own characteristically erratic way in a direction that is in many ways counter to the logic presented here. While we have tended to use separate hard structural materials and soft responsive materials, we have also tended to base our structural criteria on ensuring that the structural material itself is capable of withstanding the highest design load. This is even true for the "adaptive" aircraft wing since during take-off and landing it is the structure, and not the actuators that move the control surfaces, which withstands the primary load. We have also tended to take a very centralized approach to the design of intelligence within systems (though this is changing to some extent). Even more so, our philosophy concerning the generation of usable energy is traditionally focused on a centralized point (a power station, a compressor, or whatever) that then distributes the energy through a usually lossy distribution network. The control is via devices that are capable of withstanding considerable stresses at the point at which the energy is used. In the natural model, the energy is transported in a very convenient chemical format and controlled via the supply of the raw energy prior to its conversion into usable energy. The control system does not handle the energy that is delivered into the final load. We have managed at a very naive level to mimic some of this control, and perhaps an accelerator pedal in an automobile is the best example. Here a very gentle touch controls the flow of air into a combustion process and the results of this combustion process propel the vehicle. However, for most control systems applied to man-made machines, power-tolerant output stage control devices are essential.

At the sensing end, we have also been traditionally limited in both our approach to information gathering and our willingness and/or ability to process this information once it has been gathered. In biology, vast arrays of strategically located sensing elements with local preprocessing provide the information that drives the relatively simple actuation systems that are involved. Engineers have shied away from installing complex sensory systems within structures primarily because the vast majority of sensor devices currently in common use involve some form or other of electrical output, and the implicit wiring harness required to carry all the data from these devices to a central unit, even if the data is locally processed, becomes prohibitive. Furthermore, all the sensors in such arrays also require some form of electrical power bias. However, engineers also agree that there are many situations where comprehensive sensory data could be advantageous both to stimulate repair mechanisms and to compensate for anomalous loading.

1.7 SOME ENGINEERING LESSONS

The basic design philosophy for structural engineering has historically been to build in order to passively withstand the maximum load. This is fine for structures that see relatively modest load variation during their operational lifetime, but in many circumstances, such as buildings in earthquake zones, aircraft during take-off and landing, and vehicles in crashes, the loading can vary substantially and compromises are inevitable. A fully adaptive structure could, in principle, be configured to withstand the "normal" loading conditions in its unmodified form but to bring in appropriate actuation systems to cope with abnormal loads. This does, of course, assume that the actuation system is smaller and less expensive than the original maximum load bearing assembly.

Structures also grow old and, for example, corrosion in reinforced concrete buildings and bridges can remove the sources of their inherent strength so that even though the loads do not significantly vary during operational life, the deterioration in the structure can pose very substantial problems. The second function of interest to the engineer is the provision of an alarm to point out the need for repair processes.

There are other requirements for active responses in structures. For example, the reduction of flow noise in ships, submarines, and aircraft can provide both a quieter vehicle and one that is more energy efficient, and adaptive antenna structures may require precision positioning of parts during use to cope with such circumstances as changing thermal conditions and wind loading.

All these systems require extensive sensory input from large arrays of monitoring devices. Some signal processing, often relatively modest given the limited range of parameters to which the system must respond, is also essential. The response is furnished by an actuation system that is in many cases required to produce very high power levels over short periods and remain quiescent between times. Therefore, the engineering needs are:

- Complex multichannel sensory systems that are required for all applications;
- Sensor array interpretation and analysis, almost always requiring adaptive capabilities;
- Energy storage, conversion to mechanical work, and control within an often intermittent loading cycle;
- Precision actuation techniques monitored through the sensory array to maintain the structure within specification.

These engineering observations dictate many of the needs for research and development programs into instrumented structures. There are requirements for sensory components and array topologies; data processing and data reduction techniques; novel actuation systems; and new means of energy control, transmission, and storage. In this particular context, there is no need for the smart material—all, or much of the smartness is built into relatively standard electronic signal processing. However, there most definitely is a need for novel transduction materials and transduction material structures to optimize both the sensing and actuation processes. To many, these novel materials and novel material structures have gained the adage of being smart even though they are essentially simple transducers. The distinction is an important one even though the dividing line can become somewhat blurred unless we adhere to the information-reduction criteria, described earlier, which we shall certainly adopt throughout this book.

1.8 SMART MATERIALS REVISITED

This short description has highlighted a number of important, albeit philosophical, features that must form a part of any discussion on smart materials. Perhaps the most important is that within the definitions we have set, no single pure material could ever be construed as "smart," since all the single material can do is respond to external influences but without any implicit or explicit information-reduction potential. That is, the output signal from a single material can at best be a one-to-one function of an input stimulus; at worst it is a multivalued function of the input stimuli. Indeed, this is the rule rather than the exception since temperature almost always plays a part in the response of any single material. Any claim for material to be smart must then unavoidably necessitate the use of hybrid material mixtures configured to provide the necessary adaptive functions.

There are relatively few examples of such material combinations available at present—apart, that is, from thermal engineering and temperature compensation. There are, however, a number of very subtle examples of such thermal engineering, some commonplace. Among the most subtle and also arguably the most commonplace is an automobile temperature gauge that is designed to read "normal" over a relatively wide range but to alarm "cool" and "hot" over relatively narrow ranges through a signal that is derived from the resistance of the temperature-measuring device. Thanks to a very subtle materials engineering involving carefully controlled mixtures that are proprietary in nature, the necessary characteristics shown schematically in Figure 1.7 may be achieved. Even this smart material structure provides no control function—it simply furnishes a processed signal.

To go beyond this, we really must enter the hypothetical, and Figure 1.8 is a sketch of a hypothetical hybrid structure designed to convert an optical signal directly into a controlled power source. This could be construed as a smart material since it could participate in entirely material-based control.

These examples do illustrate that, given a functional specification, it is feasible to synthesize smart materials into structural systems engineered from a currently available set by using established rules whereby these materials interact. While it is debatable, it appears to be fundamentally impossible to fabricate such a material without recourse to some form of heterogeneous materials engineering, possibly on a cellular scale. This need should also not be confused with an equal if not greater need for new transducer materials, many of which can also be realized through similar materials engineering targeted with a view toward the transduction process rather than information reduction.

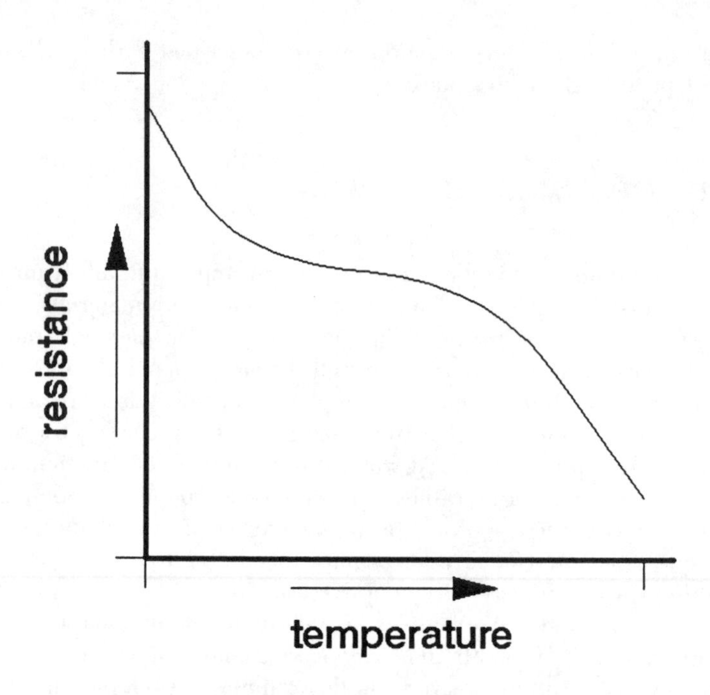

Figure 1.7 Temperature response of a "smart" resistor designed to indicate "too hot," "too cold," or "satisfactory."

1.9 HOW TO COPE WITH THIS IN A BOOK

Most texts that have sought to deal with the question of smart structures and materials have dealt with sensors, transducer materials, and actuator assemblies, treating each as a relatively isolated component. The aim in this book is to give an account of many of the materials and materials structures that are relevant to current research but throughout to contain this account within the context of the preceding somewhat philosophical discussion that has defined smart structures and materials. In essence, any activity concerned with transducer materials should be viewed as contributing an enabling technology to the hybrid smart structure that derives its sensory data, its signal processing, its actuation, and its energy resources from diverse sources. In the smart material these functions become integrated into one hybrid material structure. Indeed, at a philosophical level the difference between the smart structure and the smart material is essentially one of scale and integration rather than overall functionality indicated diagramatically in Figure 1.9.

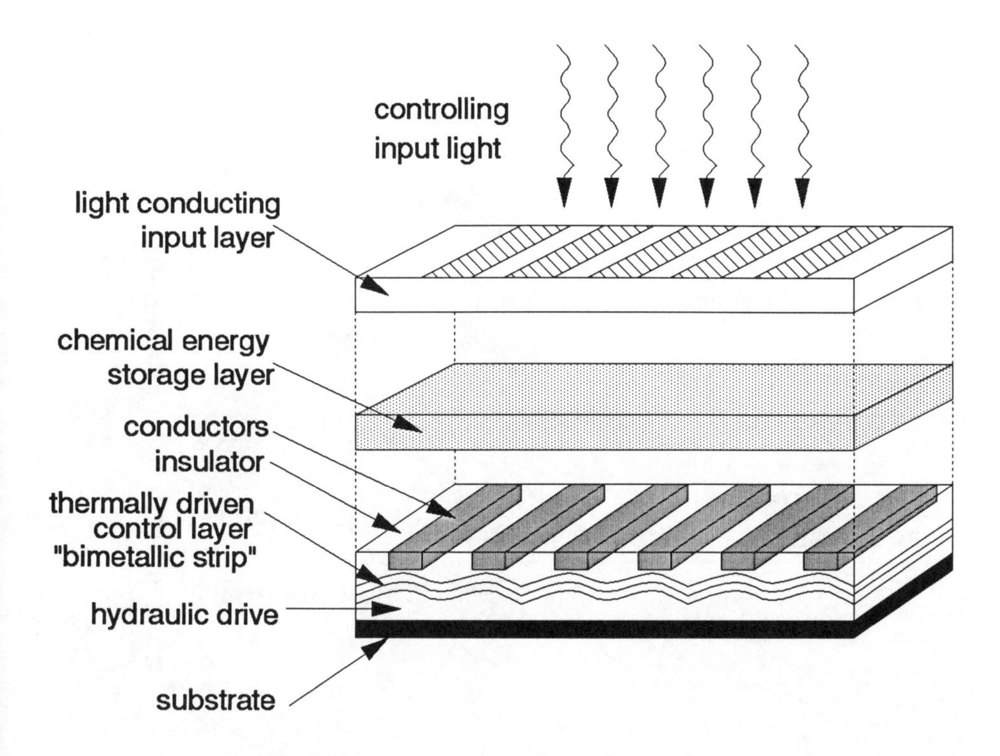

Figure 1.8 Smart "mesoscopically integrated" material for controlling significant mechanical energy using low-power optical input. Note the need for a hybrid material structure.

Within the book then we shall describe the principal features of the major sensory and actuation technologies that are available for smart structure exploitation, and we shall explore the integration issues and how these change as the scale of the intelligent unit becomes reduced. Many of these materials issues are well established in the materials literature, so within this book we shall simply highlight the major features and refer the reader to established comprehensive texts in the area. In contrast, in topics where basic research is buoyant, a more comprehensive treatment will enable the reader to achieve some perspectives on the current state of the art. The intention is to provide a balanced account of where we are, indicate what could be achieved, and speculate upon the opportunities that this presents. The hope is also to complement this with a view of the longer term possibilities offered by materials synthesis and integration, which could enable the realization of new, hence unrecognized, response functions.

The topic is genuinely highly multidisciplinary. It involves a diversity of intellects and scientific approaches, and this book must, by its very nature, reflect the experience

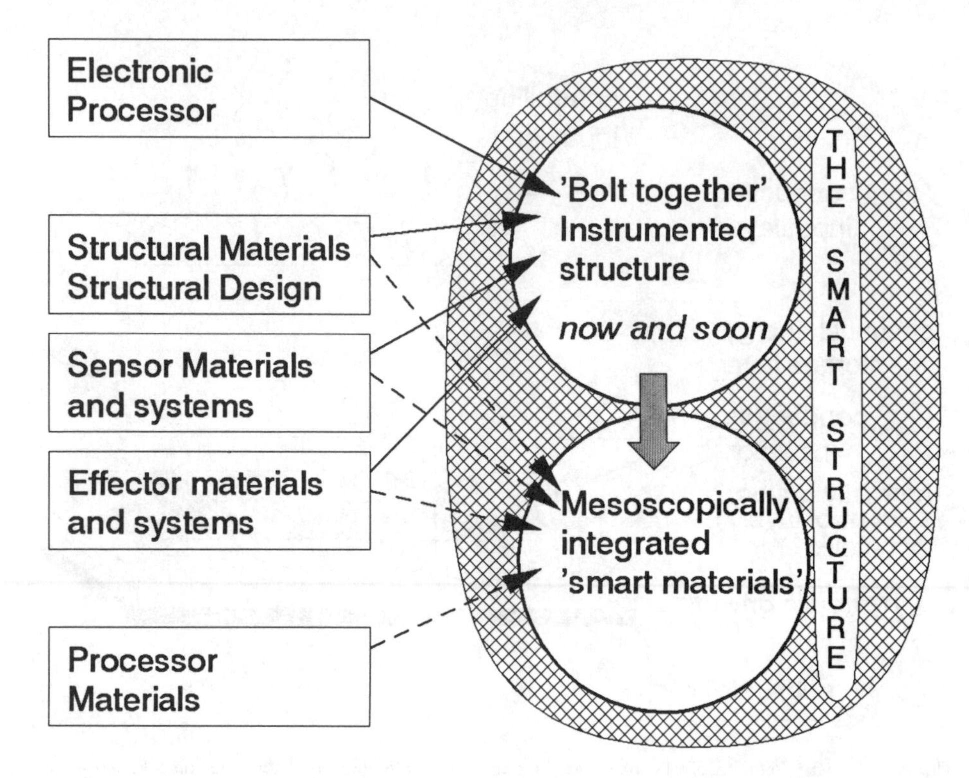

Figure 1.9 The evolution of the instrumented smart structure into the integrated smart structural material.

and predilections of its author. In preparing this text I have learned a very great deal about both the practicalities and the philosophies that underlie the whole fascinating subject of smart structures and materials. The subject is intriguing, exciting, and evolving. I hope the next few pages will convey some of the potential, the problems, and the fascination in a way that, while hopefully simple, is correct, cohesive, and self-consistent.

Chapter 2

Smart Structures—Instrumented Materials

2.1 SOME INTRODUCTORY COMMENTS

The focus in this chapter is on engineering and applications issues that dictate the potential offered by adding increasingly adaptable instrumentation to a basic engineering structure.

There are essentially two phases in the realization of this instrumented structure. The first is to attach and/or embed an appropriately devised sensing and monitoring array into the fabric of the structure. The objective here is to obtain a picture of the loading conditions and the state of deterioration of the structure. The sensor array (see Figure 2.1) must also incorporate adequate signal interpretation and data presentation equipment with the objective of presenting information to the user in the most readily assimilated form. The concept of monitoring a structure is of course far from new. Laboratory instrumentation of, for example, aircraft and spacecraft, bridges, and trains has been used for decades as a proving ground to establish confidence in a new technique or design. Many structures are monitored by inspection procedures during their operational lifetime. In principle then the concept offered by the smart structure is very similar to that familiar from laboratory test procedures. The change, which is very profound, is offered through the access to new sensing, signal processing, and data interpretation technologies, all of which are far more powerful than the equivalent of just a few years ago. This thereby eliminates the cumbersome techniques associated with traditional proof testing and provides an integrated sensing and measurement system that may be installed within a structure and from which data may be obtained continuously. The availability of such networks has a major impact upon the essential features of any design philosophy and, in particular, the following:

- Maintenance procedures can be scheduled on demand rather than at conservative intervals.

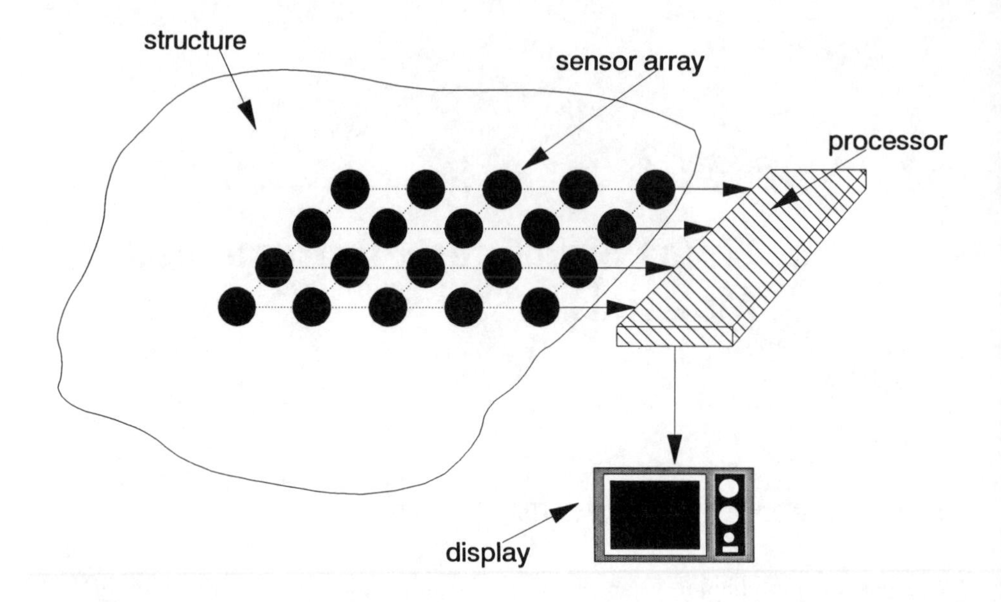

Figure 2.1 A representation of the functions of a sensor array and associated processor and display.

- Structures may be more accurately matched, at the design phase, to their loading conditions by providing for on-demand structural modifications as and when loading conditions vary.

The latter of these two points does imply some form of actuation that can be applied in response to the outputs from the monitoring system. The characteristics of this actuation can range from the exotic to the prosaic. For example, a bridge could be structurally strengthened in response to changes in legislation that permit a greater volume of traffic to pass across it, or at the more exotic end, the same bridge could be strengthened in response to load by energizing hydraulic jacks (muscular action) only when the load conditions require their presence. In this case, the foundations must still be capable of withstanding maximum load, but the structure itself can be engineered to a much lower specification. In the aircraft context, variable wing sections could replace flaps with potentially significant savings in weight (such as flap drives and mounts) and mechanical complexity.

This chapter and those immediately following will consider the broad general issues that underlie the need for active structures and will illustrate these with reference to available and emerging sensing, signal-processing, and actuation techniques. In some areas very broad applications scenarios may be painted, and these will be outlined for civil engineering and aerospace, marine, and automotive structures. The resulting needs, even from this relatively limited perspective, are wide-ranging, so the technological armory required by the smart structures engineer is diverse and far-reaching. We have limited the argument at this stage to considering purely structural applications rather than venturing into adaptive materials that may find applications in view of their electromagnetic, acoustic, chemical, biochemical, or nuclear properties. These we shall examine, albeit briefly, later in the book.

The principal observation resulting from this examination of the smart structures problem is that there is a great deal of knowledge and expertise at the components level. The uncertainties lie within issues concerning system design, system implementation, and in optimizing basic structural parameters in the light of the added dimension that the availability of advanced instrumentation systems permits. The vibration-damping community has made a few of the earlier steps in this direction. In structural engineering, which probably offers far greater potential, the implications of the availability of comprehensive and reliable monitoring systems and their impact on maintainability and structural reliability are only now beginning to be discussed. The potential offered by adaptive structural properties is barely appreciated.

2.2 THE INSTRUMENTED STRUCTURES—SOME VERY BASIC CONSIDERATIONS

The comments and observations presented in this brief section are elementary to any analysis of smart instrumented structural assemblies. Many may argue that these comments are indeed obvious, but it is surprising how often the very elementary is forgotten.

2.2.1 Functions and Responses in Instrumented Structures

The instrumented structure must:

- Monitor the physical, thermal, and chemical environment within which it is operating;
- Interpret the information derived during this monitoring process;
- Respond to changes in the physical, thermal, or environmental loading upon the structure in an appropriate fashion.

These simple observations embrace a plethora of potential sensing, processing, and response functions. In virtually all cases, the sensing function must be implemented at a large number of points distributed throughout the structure and must be capable of extracting the appropriate local measurands. The concept of load here is a very general one and can embrace vibrations, climatic stresses (for example, due to sun or wind), and corrosive deterioration in a hostile chemical environment.

The response may be twofold. The structure can either raise an alarm and seek help usually in the form of a repair or maintenance team or adapt itself to the local fluctuations that it has monitored. Usually a combination of these two responses is required.

The instrumented structure can sense and process using either active or passive techniques. The former implies the use of external energy sources (usually electrical) to sense, process, and react to information. In the latter, simple techniques, such as the release of a healing fluid from a corroded container, may be used to realize a totally passive but very limited self-repairing structure.

Active instrumentation is significantly more flexible, so the bulk of the material within this book will concentrate on this sector. However, there are very important areas where passive instrumentation can be extremely effective.

2.2.2 Structural Responses—Some Observations

Recognizing at the outset the potentially practical and distinguishing this from the essentially impractical is a critical factor in any discussion that seeks to characterize the potential offered by an instrumented structure.

As an example, the much vaunted instrumented structure that responds to unusual levels of static loading falls in the latter category. Regardless of how the structure is actually implemented, its basic framework must be capable of withstanding these unusual loading levels, so any changes in local strength must be conveyed into this framework. In most cases the adjustable-strength aspects of the structure will involve more material, more weight, and certainly more complexity than the more simple structure designed to operate under the full range of loading conditions. The 75-kg weight lifter will make no progress on lifting his or her own weight unless the bench on which he or she stands is capable of surviving the 150-kg load regardless of how adaptive or intelligent the weight lifter may be.

On this basis it seems that aircraft will continue to have to withstand thunder storms and bridges heavy traffic. However, there is considerable scope for the structure that reacts to minimize the effect of unusual external loads on its performance. A windmill is designed to turn its sails into the wind but may be configured to reduce its effective cross section to gales and hurricanes. By this means, it both optimizes its basic function and enhances its chance of survival under extreme conditions.

Less extreme examples include systems that minimize building sway in high winds, modify the reverberation time of concert halls, or actively damp out the vibration in the body panels of advanced vehicles, for example, in turbo-prop commuter aircraft.

In summary then, the responses to which an instrumented structure may have to adapt to local loading and environmental variations include:

- Minimization of the effect of unusual loading conditions;
- Reduction of structural fatigue and compensation for local environmental conditions by operations such as vibration control, modifications to wind cross section, and introducing anticorrosion reagents.

2.2.3 Sensing Systems

Sensing systems are the key to any instrumented structure and are the common element in all of them. In all cases, the sensing system must be able to resolve the parameter of interest as a function of both position and time throughout the structure. The spatial and temporal bandwidths that are required to address these measurement needs must be recognized at an early stage.

The functions of the measurement system may be divided into the following three broad categories:

- Mechanical measurements, including loading levels, resultant strains, vibration levels, and vibrational spectra;
- Thermal measurements, examining temperature distributions throughout the structure;
- Chemical and environmental measurements that are predominately designed to assess corrosion (and erosion) conditions.

The temporal and spatial frequencies with which these measurements must be made are, in all cases, inextricably linked. The time and space responses of the structure in use must be entirely determined from these measurements. With more sophisticated systems these measurements may be reduced in number significantly by using prior knowledge of the anticipated response of the structure. However, in all cases we are in the hands of the sampling theorem, which, in essence, dictates a minimum of two measurements per spatial period and two measurements per temporal period. The spatial period and the temporal period are related by the velocity of the signal of interest (in mechanical terms, usually one or other forms of acoustic displacement), so the total number of measurements on the surface of a given structure required per unit time increases as the square of the maximum temporal frequency, which is of interest in the structure.

Similar observations can be made for thermal waves in which the thermal diffusion length (broadly equivalent to the thermal wavelength) decreases as the inverse square root of the thermal input wave frequency. Again the temporal and spatial frequencies of thermal measurements are also inextricably linked, though the relationship in this case is much more dispersive than that found in the mechanical situation.

Environmental measurements are a much more difficult issue. The time constants involved are typically measured in months, but the impact of the environment on a particular structure can be extremely local. This varies depending upon imponderables such as the exact conditions under which a weld has been completed in a steel structure. The necessary spatial resolution for complete security is then extremely high though the temporal resolution can be quite coarse.

In all cases, there is a very great need for extremely careful procedures in determining the location of sensor arrays throughout a structure. The sensitivity of the array to a particular error parameter can be varied by very considerable margins by changing its relative location throughout the structure. This is one of many topics that has yet to be rigorously addressed in the problem of smart structure design procedures.

2.2.4 Self-Diagnosis

The purpose of any sensing system installed in a structure must be to enable the structure to present a record of its condition with a view to alerting some form of corrective action.

One of the most sensitive means to test any structural system is to probe it with an appropriate signal that may be, for example, acoustic, ultrasonic, optical, or x-ray and to use the interpretation of the response to this signal as a means to characterize the conditions of the system. This is the essence of all structural inspection procedures, including visual analysis. Historically, inspection procedures have been built into scheduled maintenance and have involved often expensive specialist equipment.

One potential realization of the so-called smart structure and one very powerful feature of the exploitation of the concept lies in self-diagnostic systems (Figure 2.2). In these structures a probing force (in the schematic shown in Figure 2.2 this is an ultrasonic transducer) is built into the structure itself and coupled into a system of

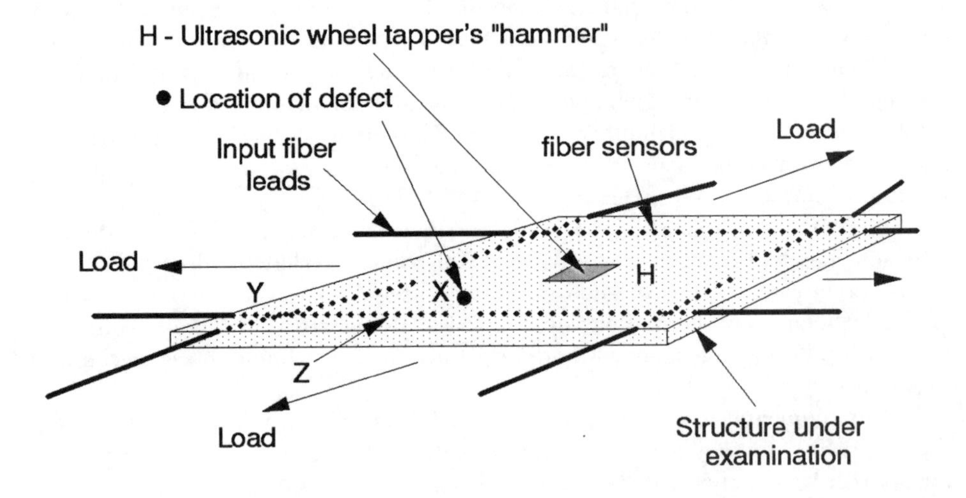

Figure 2.2 Basic approach to structurally integrated wheel tapping. The principal feature is to integrate interrogation (ultrasonic) and observations (fiber optics) into the structure and interconnect through a simple single port.

embedded sensor arrays. The response from this array of sensors to a predetermined probing signal is recorded, and changes in this response are interpreted in terms of the onset of defects, the occurrence of unusual loading conditions, or other changes in structural characteristics. The great advantage here is that the probing source and the sensing array are actually *built into* the structure itself and therefore can be addressed using a simple plug-in approach via a personal computer. This relatively simple concept—sometimes dubbed "structurally integrated wheel tapping"—promises to revolutionize routine inspection procedures and to open up the prospects for a much more structured approach to the analysis of load-bearing systems. Possible applications range from analysis of storage tanks through the monitoring of suspension components in cars and trains to automated analysis of the condition of civil engineering structures such as buildings and bridges.

2.2.5 Signal-Processing Considerations

Structural monitoring can produce quite staggering quantities of data; for example, in a structure of total volume V monitored up to a frequency f_{max} for a disturbance that travels at velocity v, the total sampling rate is of the order $16f_{max}^4 V/v^3$ corresponding to a data rate assuming a 16-bit sample of $256f_{max}^4 V/v^3$. As an example, a value of f_{max} of 20 kHz in a disturbance field, traveling at 1,000 m/s gives a data generation rate of over 40 $\text{Gbits/s}^{-1}/\text{m}^{-3}$.

While this is admittedly a high spatial and temporal bandwidth application, the general conclusion remains valid, even for much more modest performance requirements.

The data rates produced in most structural monitoring situations are so high that some form of data reduction procedures are inevitably required. This observation also links into the need for intelligent positioning of sensor arrays and the use of previous knowledge concerning the structure's intrinsic properties. There are many combinations of these basic principles together with advanced signal-processing techniques that enable reliable data interpretation. However, what is evident is that for any real-life structure these procedures are essential. This further emphasizes why the potential for smart structures has but recently emerged from the laboratory—without advanced processing techniques, much of the potential offered by state-of-the-art sensory systems is very difficult, if not impossible, to realize.

2.2.6 Actuation Systems and Effectors

The final stage in the instrumented structure is essentially the change introduced into that structure as a consequence of the decisions taken by processing the data from the sensor array. The actual format of this change can vary immensely.

At one extreme the structure can send out a request for a repair gang to come along and fix it. A gentler version of this might be to automatically spray the structure with a protective film to preclude additional corrosion or indeed to change a local electrochemical potential, thereby eliminating at least some sources of corrosion.

The mechanical response might be to energize mechanical actuators whose function is to minimize the vibration levels within the structure or to change its shape orientation such that the effects of an unusual external load are minimized. Other possibilities might include applying physical or electro-optic shutters to reflect hot sunshine, changing the electromagnetic signature to effect a camouflage function, or triggering an alarm to ensure that the human occupants vacate a building.

The principal objective of this discussion is to highlight the fact that the output from an instrumented structure can take many forms other than the directly mechanical. The discussion also highlights the fact that many instrumented structures already exist (for example, fire alarm systems) so that yet again we see that our focus lies upon the additional capability offered by combining flexible sets of sensing, signal-processing, and actuation/effector functions into unrealizable combinations.

2.3 APPLICATIONS SECTORS

It is undeniably useful to attempt to categorize the applications of instrumented structures and, from these categories, deduce general trends about the system requirements in each of the applications areas. However, simplistic classifications are thwarted by the wide range of solutions available to a particular structural problem; for example, an oil rig can be constructed of steel or concrete, and while the global features of the structure are identical in both cases, the actual monitoring requirements are very different.

Consequently the most convenient approach is first of all to examine applications by sector and secondly to establish a set of orthogonal requirements that relate to the materials from which a particular structure is fabricated.

The structures that characterize a particular applications sector may be defined in terms of the following parameters:

- The overall size of the structure;
- A characteristic dimension that indicates the distance between major structural supports within the main structure, determining, for example, the modal spectra of the structure and the maximum spatial and temporal frequency response of any instrumentation system; in very rough terms the spatial response of the system need not be more than 10 times the inverse of the characteristic dimension, and the temporal frequency response is related to this via the appropriate disturbance velocity (for estimation purposes this could be taken as 1000 m/s);

- The overall environment in which the structure must operate and within which systems within that structure must function;
- The accessibility of data from the structure and the response time of any effector action;
- The time constants of the structure to thermal, mechanical, and environmental transients;
- The frequency with which measurements need to be made upon the structure, which is in turn related to the rate at which corrective action is required;
- The costs that any additional instrumentation may add to the overall structural cost; the question as to whether the perceived benefits exceed these costs is also important, but the possibilities are much greater in structures with high value and high maintenance costs.

Table 2.1 summarizes these parameters for applications in the aerospace, civil engineering, marine, and automotive sectors. These remain the most important applications sectors, though there are also very significant possibilities in automated manufacturing, in particular in intelligent machine tools.

The materials from which structures are fabricated also have a very profound effect on the instrumentation system requirements and especially upon the form of any monitoring system. The applications sectors mentioned above use three principal types of materials: structural metals; fiber-reinforced composite materials; and concrete, frequently reinforced using steel bars. In all cases, three types of monitoring functions can be identified:

- Fabrication processes involve monitoring for the curing of resins and pastes, the setting of metals, the filling of molds, and dimensional tolerancing of machined parts. The principal areas in which new instrumentation have potential lie within the fabrication processes, especially for resin matrix composites and concretes.
- Quality assurance and testing processes must often be undertaken upon the final assembled structure and are almost always required as a part of development processes.
- In-service monitoring is essential for safety assurance and to ensure that the structure is still capable of functioning within specification.

The requirements for typical resin matrix glass and carbon fiber composite materials, for structural metals, and for concretes are summarized in Tables 2.2, 2.3, and 2.4, respectively.

These four tables present a useful summary of the features of the various applications sectors and the impact of the material choices upon system specification. While any particular application must obviously be analyzed in detail, these general guidelines illustrate both the diversity and requirements throughout the various

Table 2.1
Structural Monitoring Characteristics of Various Applications

Sector	*Size*	*Characteristic Dimension*	*Environment*	*Accessibility*	*Time Constants*	*Costs*
Aerospace	5m to 50m	0.5m	Severe	Onboard	ms-s	Can be high
Civil	100m to 1 km	10m	Moderate-severe	Can be remote	minutes, sometimes days	Moderate to high (no research tradition except Japan)
Marine	15m to 500m	1m to 5m	Moderate-severe	Can be remote	minutes	Moderate to high
Automotive	<10m	<1m	Moderate (except engine components)	Onboard	seconds	Must be low

Table 2.2
Resin Matrix Composite Materials (Includes Glass and Carbon Fiber-Reinforced Composites)

Measurand Application	*Temperature*	*Strain*	*Delamination and Fracture*	*Chemical Sensing*	*Water and Moisture*	*pH*	*Comments*
Fabrication	0 to 200°C ± 1°C (to 400°C PEEK)	20,000 $\mu\epsilon$ ± 100 $\mu\epsilon$	NA	Cross-linking of resin matrix (often indirectly via mechanical changes)	NA	NA	Cure monitoring potentially important; all sensors *must* survive cure process
Test	−55°C to +125°C ± 1°C	As above	Detect delamination ≈ 1 cm^2 Cracks ≈ μm wide	NA	Opinions vary; 0.1% moisture content sufficient to cause concern	NA	Testing primarily mechanical and thermal; moisture useful for RH testing
In-service monitoring	As above	As above	As above	NA	As above		In-service monitoring needs long life version of test sensors
Comments	Usually important to correct other measurands	Strain histories and signatures are potential indicators of structural deterioration	Can also be in principle detected by, e.g., modal frequency analysis ultrasonic testing		Effect of moisture on composites not well documented		Sensor EMBEDDING preferred and well developed, but connectors for embedded systems require development

Table 2.3
Structural Metals

Measurand Application	*Temperature*	*Strain*	*Delamination and Fracture*	*Chemical Sensing*	*Water and Moisture*	*pH*	*Comments*
Fabrication							Temperatures too high for *conventional* embedded sensors; no obvious gains from measurements Optical fiber has been embedded in aluminum
Test	–55°C + 125°C ± 1°C	1000 $\mu\epsilon$	50-μm crack widths	NA	NA	NA	Fiber optics offers simple ways to instrument complex test pieces
In-service monitoring	As above	As above	As above	Corrosion conditions and corrosion products, especially for steels	NA	Presence of pH < 9 to permit corrosion of steels	Metal (especially steel) structures also important in concrete (see Table 2.4)
Comments		Optical fiber has breaking strain 10 times that of metals	The NMI/ Cranfield tell tale one of the first in this area (≈ 1978)	Chemical may be most effective using point sensors in strategic spots for corrosion alarms			All sensors must be SURFACE MOUNTED

Table 2.4
Concretes

Measurand Application	*Temperature*	*Strain*	*Delamination and Fracture*	*Chemical Sensing*	*Water and Moisture*	*pH*	*Comments*
Fabrication	Can detect end of exothermic curing process	NA	NA	NA	Could observe drying	NA	Setting times depend on input mixture control
Test	–50°C to +80°C ± 1°C	Strain to fracture low (≈100 $\mu\epsilon$) STRESS important only in tension	Cracks >100 μm	NA	NA	NA	Structures designed to be in compression; sample testing for quality control; load models of full structure
In-service monitoring	–50°C to +80°C ± 1°C	Detect vibrational spectra or displacement under load	As above	Chloride ions could be relevant	Ingress can stimulate ice damage	pH < 10 gives signal for corrosion of reinforcement elements	Gradual lowering of pH from ≈12.8 (new concrete) to 10 in-service viewed as good indicator of potential corrosion. Erosion usually detected visually
Comments	Represents outdoor extremes	Commercial sensors already available for function. Reinforcing bars are key				Point sensors for pH on vulnerable reinforcing bars seems optimum	Building industry promising as first applications area

sectors and materials and the extent to which there are common problems that recur throughout. At this stage we have yet to discuss the appropriate technological approaches. This forms the principal feature of the following chapters.

2.4 CONCLUSIONS

The instrumented structure is currently the most important contributor to the smart structures and materials armory. While such structures have been available for decades, only the relatively recent emergence of advanced sensor arrays and high-speed digital signal processing has offered the potential for instrumented structures to emerge from the laboratory into relatively commonplace applications. The first stage in realizing such structures must be the acquisition of the necessary data from sensor arrays. The preceding sections have discussed the sensor requirements in some detail without reference to any particular technology but taking cognizance of the different applications sectors and materials options that currently dominate.

Actuators, or more generally effectors, can take diverse forms, but all must obtain the necessary information from the sensor array via an appropriate signal-processing algorithm. The data-processing capacity required in this algorithm may be very significant, and the importance of the relationship between this algorithm and the physical location of the sensor array within the structure cannot be understated. The decision from the processor should then be conveyed to a human operator or any of a wide range of electroresponsive devices and systems. The design of the actuator system must, however, recognize the practical and impractical and, in particular, should not confuse the need for *strength* required to bear unavoidable loads with the need for *adaptability* to reconfigure the structure to minimize the impact of potentially intrusive environmental perturbations.

In the chapters that follow we shall first of all discuss and describe some of the sensor technology options that are available with particular emphasis on fiber optic systems, which, for reasons that will become apparent, have gained considerable favor in the smart structures community. We shall also discuss the elements of the signal-processing requirements imposed by smart structures and continue to look at the effector options in a little more detail. However, in all cases we should recognize that the potential benefits in the application—that reduce finally to a positive cost-benefit analysis—must always be defensible. Consequently, some applications case studies will be included as examples of areas in which the benefits are real and tangible.

Chapter 3

Sensing Technologies

3.1 INTRODUCTORY COMMENTS

Most technological advances start life as a push from technologists—enthusiasts with new ideas on how to meet a technical challenge. Sooner or later, the technology meets the marketplace and adapts itself to meet real needs from which the technologists can eventually earn a living. Smart structures is no exception, though perhaps the need has existed for longer than the technology—in the early stages it was simply that technologists were not aware of the linkage.

When discussing smart structures, we must always start at the sensor end since without this the structure can give no information about its condition and so has no hope of becoming "smart." In this chapter we shall concentrate on sensor technologies. Sensing is the first and totally necessary stage in the evolution of the smart structure. The second essential is a means to interpret the data from the sensor array and thereafter to act upon this information. First and foremost, though, we must have sensors.

It is fundamentally impractical to attempt to untangle the technology from the market and the application especially since, as we have already seen, many features of the application have existed for a considerable period of time and are relatively straightforward to define. Real applications are, in the final reckoning, concerned with balance sheets; we shall leave, until Chapter 6, a discussion of how applying smart structures concepts to real situations can affect the balance sheet. Also at that stage, we shall discuss in more detail the technical specifications highlighted in the tables in the previous chapter.

This chapter will then focus upon sensor technologies and will discuss their relative merits in structural monitoring applications. We shall first discuss specifications in a little more detail and also define system-oriented terms that enable us to readily compare the potential of the various sensor technologies. We shall then discuss a number of sensor options and give a detailed account of the potential that these options offer. No discussion of this nature can be all embracing, so there will be omissions and, in time, no doubt, other important sensing techniques will evolve. Moreover, the discussion can only scratch the surface of the various techniques that are available. The intent is to explore basic comparison criteria and their limits, when these can be safely extrapolated, of the technologies that are available. To keep pace with the day-to-day evolution of smart structures techniques it is essential to review the technical literature and the conference proceedings. However, despite the apparent pace of change the fundamentals remain unscathed and new profound discoveries are very thin on the ground. To this extent the material that follows in this chapter will have, perhaps surprising, longevity.

3.2 SPECIFICATIONS AND TERMINOLOGY FOR SENSORS IN SMART STRUCTURES

We have already examined—albeit extremely briefly—the technical specifications for sensor systems in smart structures fabricated from metals, from carbon and glass fiber resin matrix composites, and from construction materials such as concrete. These requirements were summarized in the tables in the previous chapter and will be discussed in more detail in Chapter 6. However, they serve as a backdrop against which to discuss the general features of sensing systems for smart structures.

There are a number of other parameters that should be defined prior to commencing a discussion on detailed technologies. The basic *architectures* of sensor systems should be addressed, and in particular the concepts of *point sensors*, *integrating sensors*, and *distributed sensors* must be clarified:

- A *point sensor* is one that monitors a particular parameter at a closely confined point defined by the effective cross-sectional area of the sensor element. In principle, the point sensor only sees a sample of the measurand at one particular spot. In practice this spot may well extend over dimensions of the order of centimeters but rarely—essentially never—over larger dimensions. For large structures a sample of a measurand over a few square centimeters is, for all intents and purposes, a delta function representation of the parameter of interest. For many measurand fields (for example, temperature and acoustic) this point measurement may be used to infer the field over a much wider area. This area is determined by the application of the sampling theorem (if you are an engineer) or of Green's functions (if you are a physicist). For either, the basic criteria are continuity in the field of interest *and* its first derivative.
- An *integrating sensor* is one that takes an *average* value of a particular measurand over an area or length that is comparable to the area or length of the structure being monitored. Such sensors may, for example, measure an integrated strain (that is, a net elongation over a significant fraction of the structure) or an average temperature over a significant length of the structure. Integrating sensors

are sometimes spatially weighted to ensure that they are sensitized to particular spatial distributions and not to others. This concept is especially useful for modal selection in the vibration fields in large structures.

- A *distributed sensor* is capable of evaluating the parameter of interest as a function of position throughout the geometry of the sensor element. The most well-known examples exploit fiber optic technology to measure parameters such as strain and temperature as a function of position along the length of the sensing fiber. The ability to perform distributed measurements is particularly important in the structural monitoring context since it enables the derivation of the measurand at a large number of points throughout the structure using a single interrogation port and thereby eliminating the need for complex wiring harnesses. Distributed measurement technologies are, along with low-cost, high-capacity computing, one of the principal enablers in the realization of smart structures.

The principal features of point, distributed, and integrating sensors are illustrated in Figure 3.1. Sensors, again especially in the structural monitoring context, must always be assembled into systems involving large arrays of interrogation points throughout the structure. A *multiplexed* sensor system is one that combines a number of point, integrating, or distributed sensors into a complex system. The multiplexing may be implemented at several levels. It could, for example, go through an electronic interface and use techniques derived from a field bus to combine a range of electrical signals. It can also be implemented at the sensor technology level. Here the multiplexing may be effected on the measurand subcarrier. The most common example of this is the optical system that will combine the signals from several sensors onto a single optical carrier and transmit this multiplexed signal along an optical fiber before entering the decoded domain and converting the signal to an electrical one.

A half-way house is the *quasi-distributed* sensor system in which a number of integrating sensors are combined into a single system that is multiplexed in the measurand carrier domain. These concepts are illustrated in Figure 3.2.

One important parameter in structural monitoring is the need for intimate contact between the structure itself and the sensing system. The conventional wisdom of smart structures dictates that the sensors should, therefore, be *embedded* into the structure itself. A great deal of effort has therefore been expended in, for example, determining the compatibility between optical fiber sensors and structures fabricated from resin matrix composites or from concrete. While the benefits of embedding cannot be denied—regardless of the sensor technology—the process also brings with it some very significant problems. The principal one is that of interconnect (Figure 3.3). Ideally the measurand carrier should continue across any structural interfaces. In practice, this is only practical in situations where the carrier is electrical. For optical or acoustic signals there are very significant interfacing difficulties that to date have not been overcome.

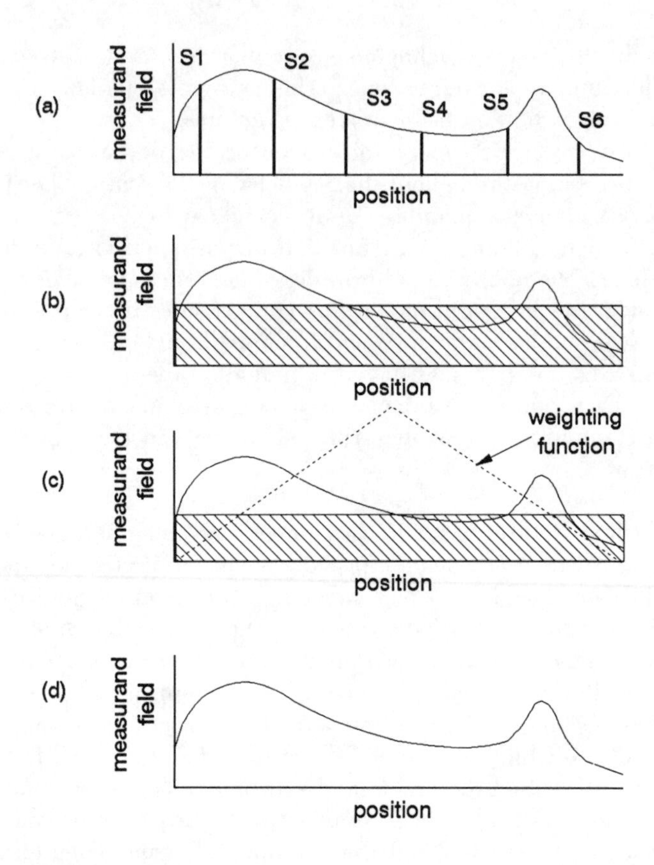

Figure 3.1 Sensor system outputs for (a) point array, (b) integrating, (c) weighted integrating, and (d) distributed sensor systems.

There are also materials compatibility issues associated with the embedding of sensors. For example, any sensor structure that is embedded in a carbon or glass fiber composite must survive the curing process and must also be able to survive the highly elastic behavior that typifies these materials that can operate at strain levels of a few percent. In concrete the sensor technology must withstand the very high pH environment, must often be pretensioned to allow for compressive loading, and for some structures must also be capable of operating in water.

Monitoring metallic structures presents a different set of challenges. Here embedding is usually totally impractical (though optical fibers have been embedded in aluminum), so everything depends upon a very efficient and reliable surface-mounting technique. Some metal structures, such as oil rigs, operate in extremely harsh environments; and surface-mounted systems are vulnerable. Further, corrosion and corrosion products should really not affect the sensors' operation.

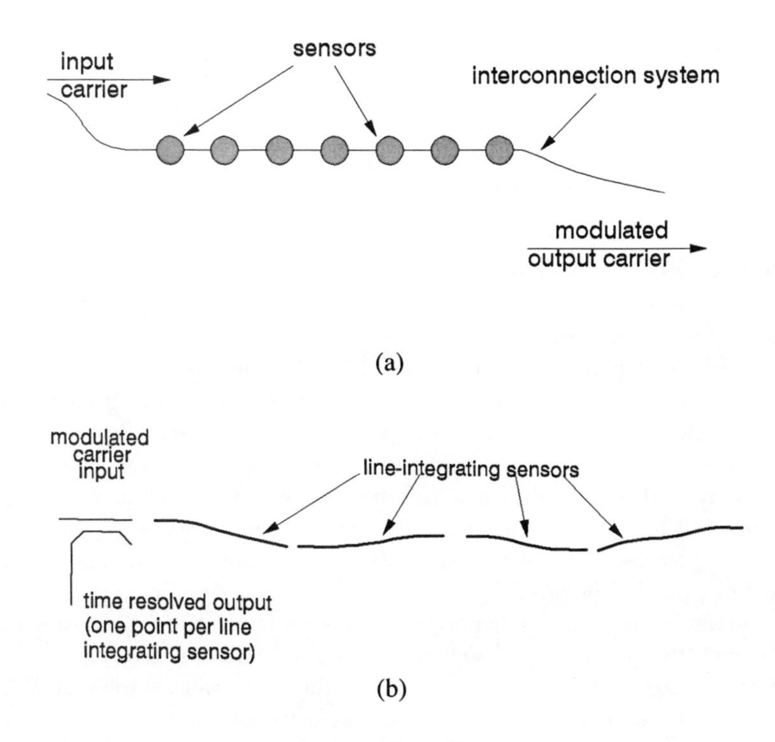

Figure 3.2 The distinction between (a) multiplexed and (b) quasi-distributed sensing systems.

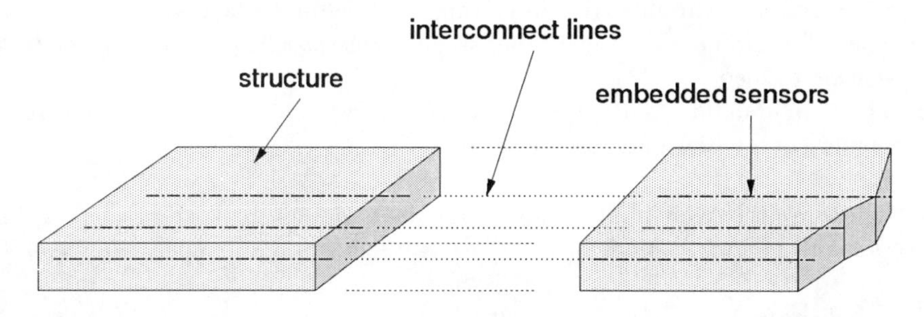

Figure 3.3 Interconnecting embedded sensor arrays require that concealed connection points operate effectively.

The requirements on a sensor system are then extremely diverse. In addition to being able to measure the parameter of interest, be it physical (typically temperature or strain) or chemical (ranging from curing of epoxies through moisture ingress to pH monitoring), the sensor must be capable of being formed into an array system to address the whole structure. It must also be compatible with the operating environment that the structure will experience. Furthermore, it should often be compatible with the fabrication of the materials from which the structure is made. With this as the overall background we can now begin to address some specific sensor technologies to arrive a comparative assessment of the potential that these technologies offer in structural monitoring applications.

3.3 SENSOR OPTIONS—AN OVERVIEW

The science and technology of measurement together present an extremely diverse combination of art, engineering, and inspiration. Sensing and measurement devices are, by their nature, extremely specialized and tailored to specific applications. The result is a fragmented technology with many small, highly specific, niche-oriented industries serving different sectors. Some idea of the extent of this fragmentation can be gleaned from the appendix in a recent book by Ohba [1] in which a list of measurement techniques extends over five pages. Any attempt to discuss sensor technologies must therefore be viewed with some caution against this backdrop of diversity since inevitably several of the potential measurement approaches will be omitted and new ideas continue to pour out of research laboratories. Sensing technologies rely upon the conversion of a carrier signal into a modulated carrier for which the modulation carries the information defining the measurand of interest. We therefore have three basic features within the sensing process:

- The process that modulates the carrier and exploits some transduction mechanism from one variable (the measurand) to a second (the carrier);
- The information carrier itself that is often also used to provide power to the sensing element;
- The demodulation of the carrier at the receiver and the transduction of the carrier energy into some usable form.

There are relatively limited choices for the variables involved in each of these processes and following Middlehoek [2] we can focus on six "signal domains":

- A radiant signal that can cover all the electromagnetic spectrum;
- A mechanical signal that includes, for example, force, position, physical dimensions, and velocity and also embraces acoustic and ultrasonic signals;
- A thermal signal that is conveniently one dimensional and defined by temperature and through that into, for example, thermal capacity and thermal resistance;
- An electrical signal that embraces, for example, voltage, current, and electric field;
- A magnetic signal that is defined through the magnetic field (or the magnetic flux density as appropriate); while this signal is always directly related to an

equivalent electrical signal, the two are conveniently separated since the "molecular currents" that give rise to ferromagnetism are not conveniently recognized separately;

- A chemical signal domain that defines the composition of matter and parameters such as crystal structure.

These signal domains do clearly overlap one with the next, but the classification is useful and graphically illustrates the diversity of sensing technology. In Figure 3.4 we show a representation of this based upon Middlehoek's sensor cube. In each elemental box within the cube we can define some complete transduction system. In many of these elemental boxes there is a wide range of physical and/or chemical phenomena that can be used to fulfill the transduction process represented by the box. Indeed, the function of this diagram is primarily to emphasize very graphically the diversity of sensor technologies while at the same time allowing some common features to be derived.

In the context of smart structures, the focus must be on measurements of a relatively limited range of physical and chemical parameters and also embrace the need for an *electrical* signal output since it is this domain in which the necessary signal-processing capabilities are available. In other words, our *output* signal must lie in the electrical plane, but the input signal applied to the modulating channel can exist anywhere within the 36 options inside that plane. Table 3.1 looks at the five

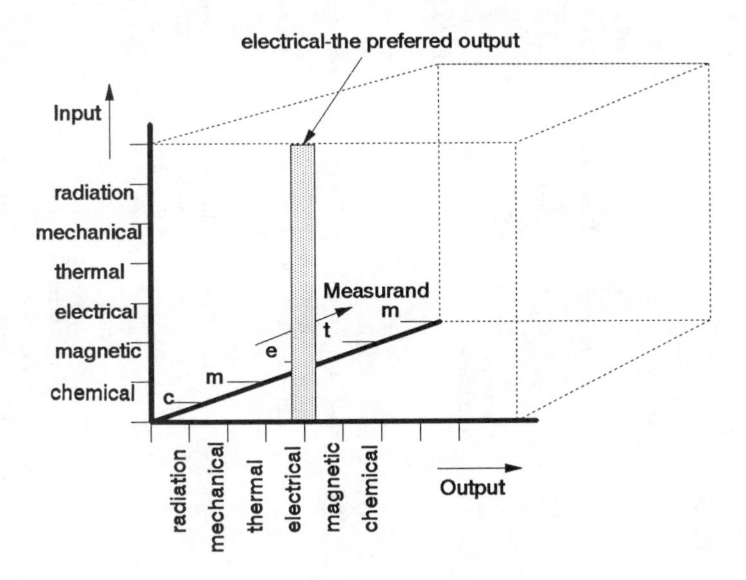

Figure 3.4 The sensor cube—a graphic indicator of the very wide range of techniques that are available for measurement.

Table 3.1
Some Transduction Mechanisms: All Can Form Sensors
(There Are Many More)

Input Variable	*Output Variables*								
	Current	*Voltage*	*Resistance*	*Light*	*Temperature*	*Magnetic Field*	*Pressure*	*Strain*	*Chemical*
Current		Electronic Amplifiers	Thyristor	Light bulb	Joule heating	Direct relationship/ Ampère's Law	Via e-m forces	Magnetostatic forces	Electrolysis
Voltage				Gas discharge	Resistive/ Joule heating	via current Ampère's Law	via e-m forces	Electrostatic forces	Ionization potentials
Resistance	Ohm's Law	Ohm's Law							
Light	Photoelectric effect	Photovoltaic effect	Photo conductivity	Fluorescence	Absorption Bolometry		Radiation pressure	Radiation pressure	Photo etching photosynthesis
Temperature		Thermocouple	Resistance thermometer	Pyrometers			Thermally induced stresses	Thermal expansion	Thermal decomposition
Magnetic field	Faraday's Law/Induction	Hall effect	Magneto-resistance	Birefringence in domains	Dynamic hysteresis losses		Magnetic forces	Magneto-striction	
Pressure		Piezo-electricity	Piezo-resistance	Piezoelectric breakdown	Dynamic stressing	Magneto-striction	Fluidic amplifiers	Diaphragm motion	Phase changes
Strain		Piezo-electricity	Piezo-resistance		Dynamic straining	Magneto-striction	Direct displacement	Levers	
Chemical	Ionic concentrator	Electro-chemical cells	Chemi-sorption	Chemi-luminescence	Heat of reaction		Volume changes in reaction	Volume/heat of reaction	All chemical reactions

nonelectrical signal domains and defines a range of phenomena that will translate from one of these nonelectrical domains into electrical output. Again this simply indicates the breadth of the sensor options, but also remember that this only represents one plane within the sensor cube. These phenomena in Table 3.1 could also act as carriers for information from other transducers.

The conclusion from this preliminary discussion is simply that this fragmentation of sensor technology defies any attempts at simplistic classification. The consequence is that any discussion of sensors that assesses the options available must be driven by an applications focus and can never be all-embracing. Technological advances also play a very important part, and in particular the power of signal processing and data extraction is immense and has already had a profound effect on sensor technology.

The broad requirements for smart structures were defined in Chapter 2. If we concentrate first upon mechanical (that is, strain and displacement measurements), then the principal options that are currently available are shown in Table 3.2. The techniques in this table are divided into two broad categories. The first six columns involve modulating an electrical carrier and therefore by implication require a wiring harness throughout the structure, the second two utilize an optical carrier and, in effect, have substituted for the wiring harness increasing degrees of sophistication in the signal processing in order to achieve precision measurements. Some of these options are discussed in more detail later in this chapter.

Temperature measurements are also important in structural monitoring. Often the temperature measurement is required simply to correct the value of, for example, a physical measurement, but it can also be used to indicate thermally induced stresses and to give a guide on the speed on which corrosion processes are proceeding. Table 3.3 shows some of the principal techniques for thermal measurement and again there is a distinct difference in character between the methods that use an electrical communication channel and those that use an optical one.

Many smart structures are inherently large, extending over tens or even hundreds of meters. In such circumstances the optical carrier is potentially very attractive—its immunity to crosstalk and corrosion resistance are very attractive. This chapter will feature extensive coverage of optically based measurement systems, reflecting not only my own research interests but also the very real applications benefits that these systems afford.

It is paradoxical that in many cases—some would argue most—the real parameter of interest is chemical activity, giving an indication of corrosion and decay processes within the structure. Often the strain, displacement, and temperature measurements are used as secondary indicators of structural changes that have been induced through some sort of chemical activity. For most physical structures this chemical activity involves the presence of moisture together with other indicators such as pH or oxygen presence. Some of the principal chemical-measuring systems that may be applied to

Table 3.2
Strain and Displacement Sensors: The Principal Options for Structural Monitoring

Type	*Foil*	*Semiconductor*	*LVDT*	*Vibrating Wire*	*PZT Film*	*PZT Ceramic*	*Optical Fiber*	*Photogrammetry*	*Comments*
Typical Sensitivity	30 V/ϵ	1000 V/ϵ	~1 V/fsd	x 2 in frequency for 0.1%ϵ	10,000 V/ϵ	20,000 V/ϵ	2 fringes/μ 10^{-8} fringe detectable	100 $\mu\epsilon$ or better	10 $\mu\epsilon$ is typical structural need
Point/Integrated	Point (integrated to 0.1m)	Point	Displacement	Point over short gauge length	Can integrate using large films	Point	Can be either	Surface or volume measurement	Relative merits of point and integrated sensing complex: see text
Gauge Lengths	0.5 mm to 100 mm	1 mm^2	0.5 mm to 600 mm (displacement)	From millimeters to centimeters	From millimeters to a few meters	From millimeters to centimeters	From millimeters to hundreds of meters	To a few meters	The necessary gauge length can be gauged from Table 2.1 (characteristic dimension)
Bandwidth	~10 kHz	~100 kHz	<100 Hz	~100 Hz	AC (few Hertz) to 100 MHz	AC (few Hz) to 100 MHz	DC to 100 MHz	Essentially DC because of processing need	See also Table 2.1 and comments on structural dimensions; high values needed for ultrasonic probes
Distributed measurement potential	No	No	No	No	Possible, not demonstrated, and requires multiple connections	No	Yes—relatively straightforward	Inherent but needs markers	Distributed measurement capability is virtually essential in on-line, continuous structural monitoring; for test and evaluation, can tolerate using wiring harnesses

Table 3.2 (continued)

Type	*Foil*	*Semiconductor*	*LVDT*	*Vibrating Wire*	*PZT Film*	*PZT Ceramic*	*Optical Fiber*	*Photogrammetry*	*Comments*
Multiplexing feasibility	Difficult: field bus	Difficult: field bus	Difficult: field bus	Difficult: field bus	Can be integrated into film layout. Readout difficult not demonstrated	Difficult: field bus	Can be integrated into fiber design. Readout demonstrated for many systems	Not applicable	
Chemical material compatibility	Poor	Good/can be passivated	Poor	Poor but can be packaged and has been used in concrete	Poor—low temperature Curie point	Poor brittle	Excellent, withstands many processes	Inherent because noncontact	See Tables 2.2 to 2.4 for summaries of chemical requirements
Dynamic Range	1.0%ϵ	to 0.5%ϵ	0.3% of f.s.d	~0.1%ϵ	~0.5%ϵ	~0.1%ϵ	to 5%ϵ only useful in tension	Not applicable	Large dynamic ranges needed in, e.g., composites. Some materials, e.g., concrete often operate in compression
Comments	The traditional strain gauge	Excellent for point measurements	Excellent, well-characterized displacement measuring device, though large	Already in use to monitor some concrete structures. VWG's are large (relatively)	Low Curie temperature restricts application (80°C). Excellent potential as distributed gauge		For many, the most promising technology. Distributed capability very powerful	Relatively little applied to date. Good for inspection if define markers on structure and solve complex processing issues	

Table 3.3
Temperature Measurement

Type *Parameter*	*Thermocouple*	*Platinum Resistance*	*Thermistor*	*Displacement Gauges*	*Optical Fiber (Raman)*	*Pyrometry*	*InfraRed Imaging*
Range	To 1800°C depends on materials forming junction	to ~700°C	to ~300°C	Typically –50 to +150°C	–50 to 200°C (Depends on fiber coating survivability)	>500°C to 4000°C	–50 to +100°C
Resolution	<0.2°C	<0.2°C	Can be <0.02°C	~1°C (to 0.1°C)	~1°C on 1 meter of fiber path	~1°C to 5°C	Can be <0.05°C
Stability	Depends on reference junction	Determined by bridge components	Depends on resistance measurement	Depends on mechanical construction	Uses Raman effect, inherently stable	Depends on reference and emissivity	Usually *relative* measurement over image field
Cost	Low, but needs interconnect	Low	Low	Low	Very high	Medium	Medium to high
Communication Channel	DC electrical voltage	DC electrical resistance	DC electrical resistance	Visual/ mechanical	Optical	Optical	Optical imaging
Comments	The principal industrial temperature measuring system. Very well understood and characterized. Can be used over long distances	Used for reference thermometry. Self-heating and lead effects can be problematic. Only used over relatively short distances	Can be very small. Good for "local" measurements	Mercury in glass most common. Bimetallic strips also very common	Enables *distributed* measurements. Only commercial system to do so	Only usable simple system for temperatures >1000°C	Excellent for analyzing, e.g., two-dimensional thermal fields, locating hot spots

these circumstances are shown in Table 3.4, and again the options that are the most important will be explored in more detail later in this chapter.

3.4 PHYSICAL MEASUREMENTS

Physical measurements—that is, those of strain, displacement, and temperature—are currently among the most important required in smart structures applications. There is an enormous amount of literature describing the techniques of physical measurement and a gradually evolving research activity that is ongoing in most of them. In the following sections we outline the underlying principles of the major contributing technologies used in structural monitoring and discuss their relative properties.

3.4.1 Piezoelectric Strain Measurement

The piezoelectric effect is very simply described in terms of the ability of a range of materials to convert mechanical stress into electric field and vice versa. The use of the effect in sensors obviously depends upon the latter property.

In order for the piezoelectric effect to occur, the materials concerned must be anisotropic and electrically poled, that is, there must be a spontaneous electric field in a particular direction maintained throughout the material [3]. It is this spontaneous field that disappears above the Curie point, which distinguishes piezoelectric materials from electrostrictives. While piezoelectrics have applications for both strain measurement and actuation, electrostrictives are limited to the actuation function. This will be discussed in Section 4.4.

Conventionally, the properties of piezoelectric fields are referred to this polarizing axis P (see Figure 3.5), which is always referred to as the "3" axes [4].

The responses of piezoelectric crystal are determined by the plane stresses applied in the directions 1, 2, and 3 and the shear stresses applied in the directions 4, 5, and 6, which are perpendicular to the axes 1, 2, and 3, respectively. The piezoelectric material is an anisotropic crystalline structure, so the general response function is quite complex though the principles themselves are relatively straightforward.

The response of the piezoelectric material is determined by

$$\mathbf{D} = d\mathbf{X} + \epsilon^{x}\epsilon_0\mathbf{E} \tag{3.1}$$

$$x = s^{E}\mathbf{X} + d\mathbf{E} \tag{3.2}$$

where $\mathbf{D}$ is the electric displacement vector, $\mathbf{E}$ is the electric field, ϵ^{x} is the relative dielectric constant at a particular value of $\mathbf{X}$, $\mathbf{X}$ is the stress applied to the material,

Table 3.4
Chemical Measurement Techniques[1]

Technique Parameter	*Spectroscopy**		*Chromatography*	*Selective Absorbers*	
	Linear absorption	Nonlinear (e.g., Raman and Fluorescence)†		Chemical: Mechanical converters	Chemical: Electrical converters
Selectivity	Inherently high but needs good knowledge of absorption spectra, which sometimes overlap	Nonlinear processes often temperature dependent. Raman spectrum can be broad and ambiguous. Fluorescence often suffers cross sensitivity since uses indirect chemistry	Liquid and gas phase systems well characterized and highly selective	Generally poor, but some (e.g., Pt for H_2) very selective‡	Depends on chemistry. Generally measuring ion concentration
Stability	Inherently high. Refers to spectrometer	For Raman depends on spectrometer and temperature gives intensity. Fluorophores decay and can be poisoned	Very good—effectively refers to a time reference	Varies from very high to poor depending on transducer material	Fair to good. Most have usable life from hours to years

[1]Solids can usually only be probed at the surface, otherwise they must be dissolved. Most chemical analysis techniques address liquids and gases.
*Spectroscopy can be made very cost effective by tailoring a spectrometer to particular applications.
†Many nonlinear processes, e.g., fluorescence, operate through intermediate chemistry.
‡A very wide range of materials exhibit this type of behavior, especially in the presence of moisture, which is a most important structural monitoring parameter.

Table 3.4 (continued)

Technique Parameter	*Spectroscopy**		*Chromatography*	*Selective Absorbers*	
Comments	Simple and highly selective. Some species (e.g., hydrocarbons) require sources in near IR where there are few that operate at correct wavelength. Absorbance depends on concentration of species of interest and is essentially independent of other parameters. Best suited for gas analysis. Poor in solids	Usually need high input power densities to excite nonlinear processes. Some processes nonreversible. Signal frequency shifted from excitation and thus immune to, e.g., reflections and scatter. Can be used in solids (for surfaces). Best in liquids. Gases have low cross section	Only really suited to analyzing laboratory samples. Unsuited to field applications	Relatively little exploited in measurement though some (e.g., drug release valves) very powerful	A very wide variety of chemical systems in this category; especially for gas sensing, for, e.g., enzyme-based electronics

[1]Solids can usually only be probed at the surface, otherwise they must be dissolved. Most chemical analysis techniques address liquids and gases.
*Spectroscopy can be made very cost effective by tailoring a spectrometer to particular applications.
†Many nonlinear processes, e.g., fluorescence, operate through intermediate chemistry.
‡A very wide range of materials exhibit this type of behavior, especially in the presence of moisture, which is a most important structural monitoring parameter.

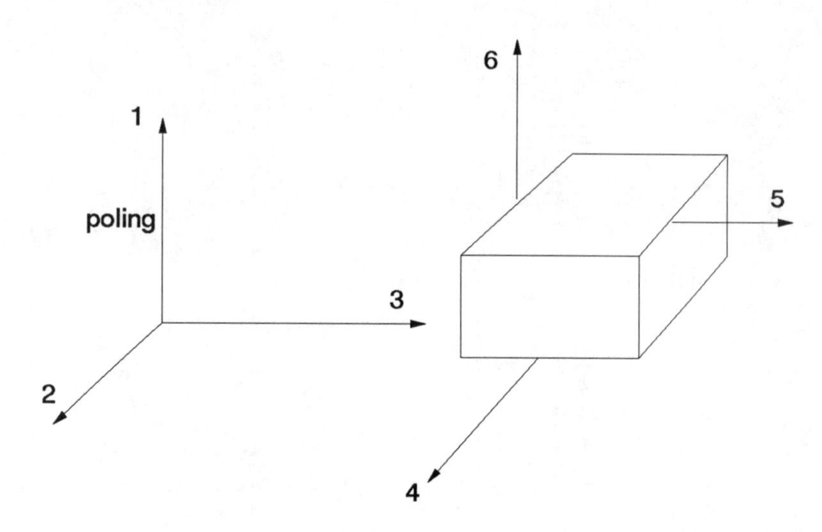

Figure 3.5 Principal axes for piezoelectric systems defining stress directions (1, 2, 3) and shear planes (4, 5, 6).

x is the strain, s_E is the mechanical compliance at a particular value of E, and d is the *piezoelectric strain coefficient.*

In general the quantities **D**, **X**, and **E** are vectors and d, s_E, and ϵ_x are tensors.

If we limit the discussion to poled polycrystalline materials that includes most practical applications, then only d_{33}, d_{31}, and d_{15} are required to define the structural response together with the five elements of the strain tensor s_{11}, s_{12}, s_{13}, s_{33}, and s_{44}.

From the point of view of strain sensing the necessary conversion between the applied strain x and the consequent stress X can be obtained through (3.2). The actual *voltage* produced by the strain field is obtained by integrating E along the appropriate direction. The value obtained also depends on the electrical load, so it is convenient to compare the performance of piezoelectric sensors in terms of the *open circuit* voltage response to a predetermined strain. The open circuit response is equivalent to operating under a constant value of D—that is, the *charge* on the surface of the piezoelectric material remains constant.

Consequently, by rearranging (3.1) and (3.2) we can derive the open circuit electric field E_0 as a function of a dynamic applied strain x by

$$x = \left[d - \frac{s^E \epsilon^x \epsilon_0}{d} \right] \mathbf{E} \tag{3.3}$$

Remember that d, s^E, ϵ^x are tensors, so this simple equation conceals significant complexity!

The other major factor that determines the response of a piezoelectric material is the electromechanical coupling constant defined by

$$k^2 = \frac{\text{Mechanical energy converted to electrical energy}}{\text{input mechanical energy}}$$
$$= \frac{\text{electrical energy converted to mechanical energy}}{\text{input electrical energy}} \quad (3.4)$$

The behavior of k^2 can be derived from the equivalent circuit for a high Q factor piezoelectric element shown in Figure 3.6 in which C_0 is the clamped capacitance of the piezoelectric element and L_1, R_1, and C_1 represent the electromechanical conversion processes. It can be shown by a lengthy but straightforward process that

$$k^2 \approx \frac{C_1}{C_0 + C_1} \approx \frac{f_p^2 - f_s^2}{f_p^2} \quad (3.5)$$

where f_p and f_s represent parallel and series resonant frequencies, assumed close to each other as indicated in Figure 3.7.

In structural sensing applications the piezoelectric system is almost invariably used at frequencies well below resonance (except for ultrasonic receivers). Under these conditions the impedance is dominated by the capacitance terms in Figure 3.6. Ideally then the amplifier into which the piezoelectric transducer is interfaced should have this high matching impedance. Remember that in this case the boundary conditions will inhibit the effective value of E and will in turn conspire to reduce the achievable D, which determines the voltage input to the amplifier. For comparative purposes it is useful to discuss the *open circuit* response of piezoelectric strain gauge transducers, which is equivalent to specifying a charge amplifier.

The above equations are adequate to specify the electromechanical response of piezoelectric materials. While the details are often quite complex, the principle is

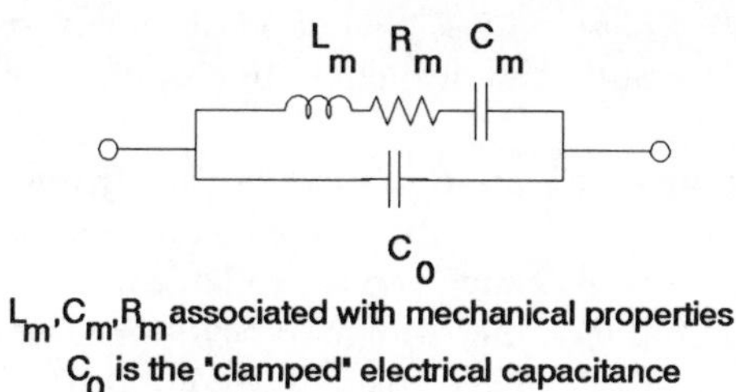

Figure 3.6 Electrical equivalent circuit for piezoelectric plate operating near resonance—used to derive electromechanical coupling coefficient.

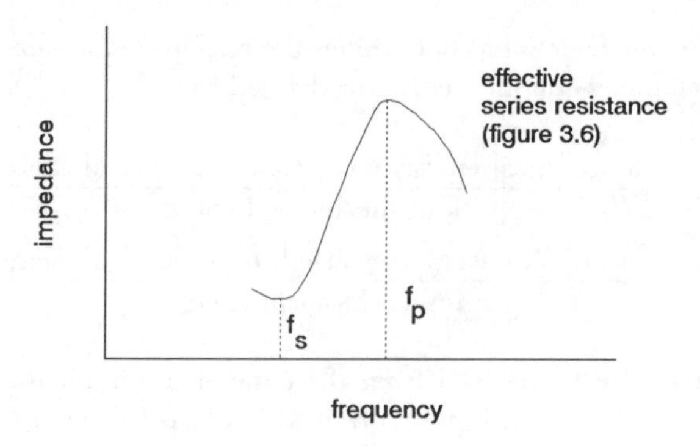

Figure 3.7 Measurement of f_p and f_s to determine the electromechanical coupling coefficient.

very simple—ensure that both mechanical and electrical impedances are matched throughout.

There are two principal types of piezoelectric materials: ceramics and polymers. Table 3.5 summarizes the principal properties of some commonly used piezoelectric materials. The table concentrates on piezoelectrics used in strain measurement and actuation. One important material—crystalline quartz—has been omitted since the principal function of this material is as an electromechanical oscillator with an inherently extremely high mechanical Q factor (exceeding 10^6). The role of this material as a reference oscillator is well known but is outside the scope of the present discussion.

Piezoelectric ceramics, which are probably the most important piezoelectric materials, may be thought of as comprising randomly oriented piezoelectric crystals. Indeed the ceramic materials are fabricated from components that exhibit piezoelectricity in their crystalline form. The principal feature of piezoelectric crystals is the inherent requirement for some form of asymmetry within the unit crystal cell. Most materials are, or are derived from, the so-called peroskite structure (see Figure 3.8). The solid-state textbooks invariably insist upon using barium tantalate as the model, and while the principles do apply, most practical systems are based upon lead zirconate/lead titinate solid solutions (though occasionally single-crystal materials such as lithium tantalate and lithium niobate find applications in the piezoelectric mode).

Single-crystal piezoelectric materials can obviously only be polarized down a particular axis and indeed often exhibit spontaneous polarization. They are then effectively uniaxial and are only conveniently available in the form of relatively small platelets. Their applications tend therefore to be as mechanical oscillators (the quartz crystal) or in surface acoustic wave devices. They are little used as sensors though there are a few occasions where their inherent anisotropy can be used to good effect to separate the influences of strains in one direction from those in another, for example, in accelerometers.

Table 3.5
Properties of Piezoelectric Materials

Property	*Units*	*Ceramics*				β phase Polymer PVDF (typical)
		*PZTA**	*PZTB**	$Na^{1}/2K^{1}/2NbO_3$	$LiNbO_3$†	
Density	$\times 10^3$ kgm m^{-3}	7.9	7.7	4.5	4.64	1.78
Curie Temperature	°C	315	220	420‡	1210	100
Young's Module (Y_{11})§	Gpa	~70	~70	~100	~100	2
Maximum Strain	μe	1000	1000	1000	1000	700
d_{31}		–119	–234	–50	–0.85	+20
d_{33}	pC.N^{-1}	268	480	160	160	–33
d_{15}		335	—	—	—	—
k_{31}	.33	.39	.27	.02	.12	
k_{33}	.68	.72	.53	.17	.19	
k_{15}	.68	.65	—	.61	—	
ϵ_s		1200	2800	500	29/85*	12
Breakdown Field	Vμm^{-1}	~1	~1	~1	~1	~100

*PZTA and PZTB are indicative of many Lead Zirconate Titanate ceramic formulations.
†Single-crystal material.
‡Loses polarization at ~ 180°C.
*This is effectively one element of the inverted compliance tensor S_E.
§$\epsilon_{33}/11$.

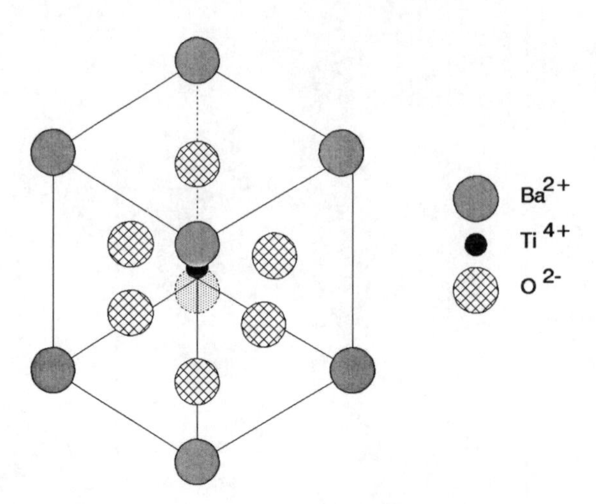

Figure 3.8 The Perovskite structure, common in ferroelectrics. Below the Curie temperature, the oxygen ions are displaced below the crystal center, resulting in a net dipole moment.

The vast majority of ceramic piezoelectrics are in the PZT family. Figure 3.9 shows the phase diagram for lead zironate titanate system [5]. The principal reason for presenting this diagram is to indicate that even the simple system has a wide variety of possible properties and that the addition of proprietary dopants introduces yet more variants. PZT-based ceramics have two principal benefits:

- They furnish materials with a high piezoelectric activity.
- These materials can be molded during their preparation into a wide variety of shapes, and, of these, plates and hollow cylinders are particularly useful. The ceramic processing also enables the fabrication of piezoelectric stacks (Figure 3.10).

The ceramic is inherently isotropic so that the polarization axis may be determined by the polarization process itself; for example, radial polarization of the hollow cylinder is very simple to arrange. The polarization processing involves applying an electric field across the sample in excess of the cohesive field of the material (Figure 3.11), which is typically of the order of 1 MV/m, raising the temperature of the material above the Curie point and cooling the material below this point to "lock" the domain structure.

Piezoelectric ceramics are extensively used as strain sensors (equivalently as acoustic detectors) and as mechanical sources (see Chapter 4) for both displacement drives and acoustic sources in air (high-frequency tweeter systems) and in water where they can generate very substantial acoustic power densities.

We have already mentioned the need for careful impedance matching both at the mechanical and electrical sides of the circuit. For many applications PZT-based

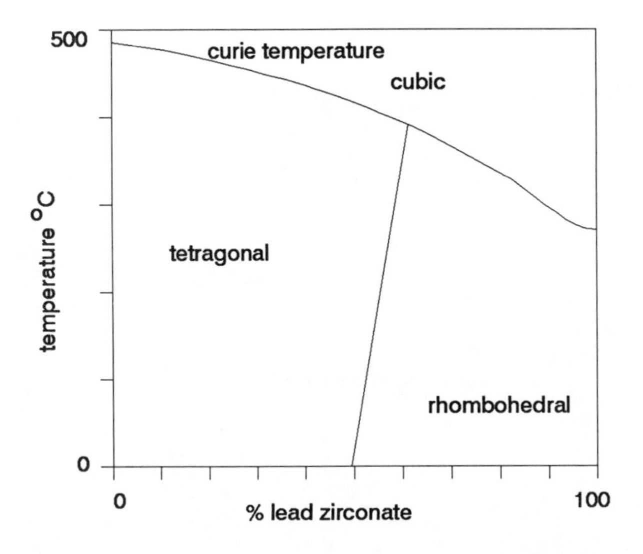

Figure 3.9 Simplified phase diagram for lead zirconate/titanate system, indicating changes in crystal lattice and curie temperature.

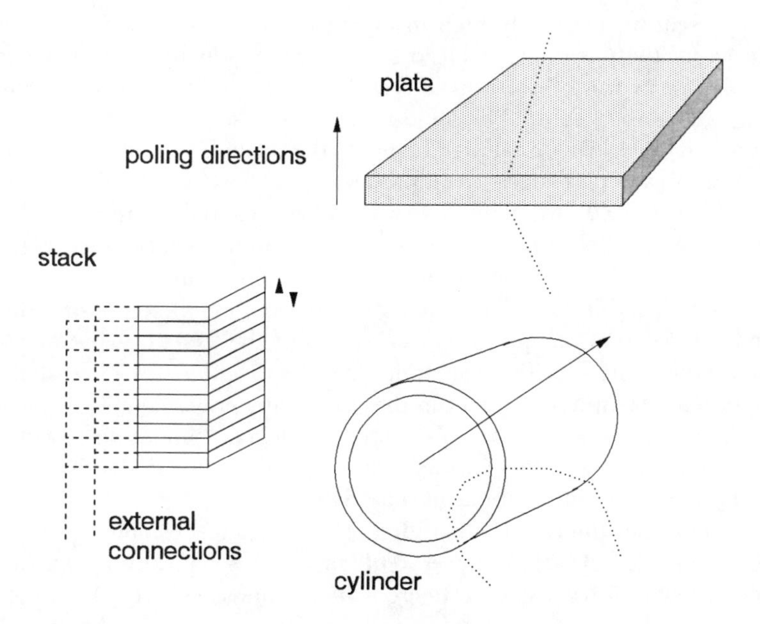

Figure 3.10 Some possible shapes for piezoelectric elements.

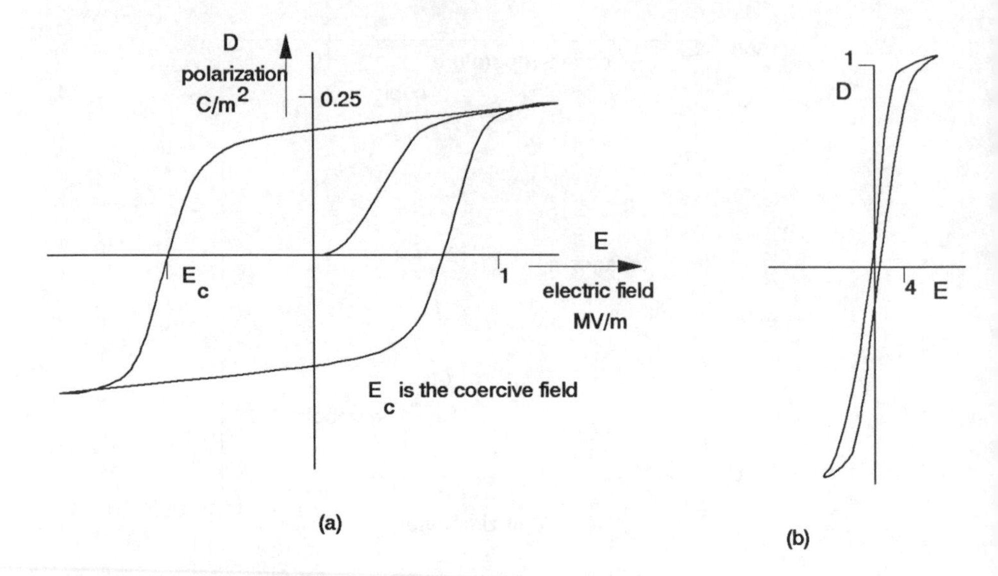

Figure 3.11 The D-E loop for (a) piezoelectrics and (b) electrostrictives. The latter exhibit much lower hysteresis losses.

ceramics present an extremely high mechanical impedance and for certain strain fields their sensitivity is much reduced by the way in which the d coefficients add; for example, for hydrostatic strains the resulting electric field depends primarily on $d_{33} + 2d_{31}$ and these (Table 3.5) almost cancel. Consequently there is a need for careful material and structure design to match the strain field to the transducer. This leads into the area of piezoelectric composites [6] in which conventional piezoelectric materials such as PZT are combined with other materials—often a polymer but including the possibility of honeycomb structures and a host of others (Figure 3.12) in order to furnish the appropriate combination of impedance matching and strain field distribution sensitivity. The detailed implementation of composite transducers is beyond the scope of our present discussion, but the basic concept of producing material mixtures with specific orientations in order to optimize the overall properties of the material system lies well within the "smart materials" remit. These material mixtures are effectively structurally based analogue computing systems "programmed" to perform a particular mechanical (or other) function as well as possible by using appropriate combinations of materials.

Polyvinylidene fluoride (PVF_2), or (CH_2-$CF_{2(n)}$), has established itself as the most important piezoelectric polymer. This material is typically available in sheet form from a few to a few hundred microns in thickness. The very high poling field (of the order of 100 V/μm) in effect limits the maximum thickness that is realistically available. The material is prepared by first mechanically stretching material obtained

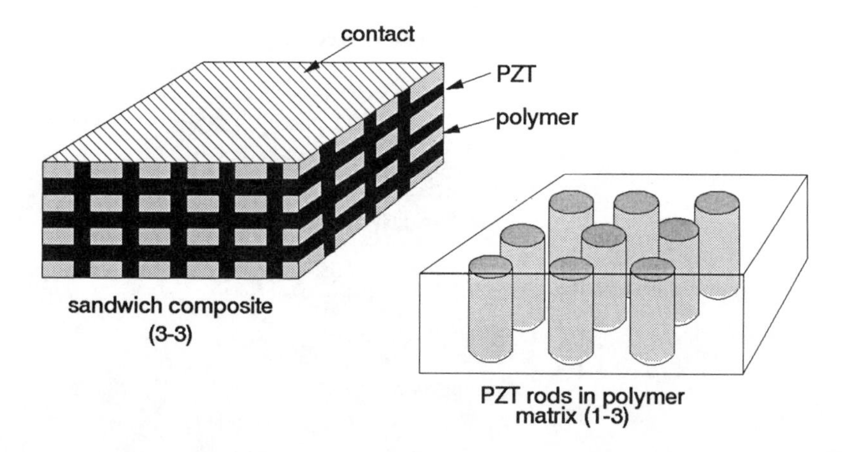

Figure 3.12 Piezoelectric composites for impedance matching. The transverse dimensions of the composite cell are generally ≪ wavelength.

from a melt and then applying the field transverse to the stretching direction (Figure 3.13) [6]. The principal applications features of PVF_2 are:

- It is available in a very easily handled sheet format at low cost.
- The polarization direction is always in the thickness direction of the sheet.

PVF_2 is an attractive sensing material since it is easily handled and has a good strain-to-voltage conversion efficiency. It may also be configured in very simply varied geometrical formats. Against this it has a relatively low Curie temperature (typically of the order of 100°C) and can be difficult to pole and the piezoelectric coefficients have very significant temperature coefficients over normal ambient operating ranges.

There are a few piezoelectric materials that may be processed in thin-film format and therefore offer an additional fabrication flexibility. Of these the most commonly encountered is zinc oxide [7], which, with care, can be sputtered in crystalline form onto a variety of substrate materials, in particular silica. This material combination offers some potential for both sensing (since the transducer can be built into an integrated circuit) and actuation, though to date it has been but little exploited since sensing on a silicon substrate may be readily implemented using other means, (piezoresistivity) and the actuation forces are relatively small.

The great majority of sensing applications based upon piezoelectrics use either piezoceramic or polymer transducers. Both these materials are essentially isotropic. The polarization axis is defined by the polarizing field (that is, the "3" axis). Even with this relatively simple system, great care is required in the use of piezoelectric transducers and the interpretation of the consequential results.

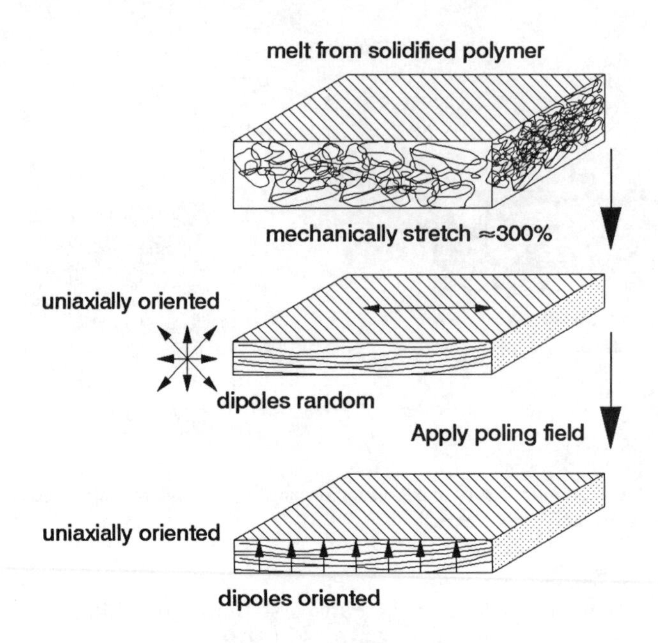

Figure 3.13 Preparation stages of a piezoelectric polymer.

There are essentially two modes of operation: The strain is applied along the "1" axis (that is, in any direction perpendicular to the "3" axis), and the resultant voltage measured through the electric field change along the "3" axis or the strain may be applied along the "3" axis and the measured voltage along the same axis.

In principle, the appropriate behavior can be derived from the equation

$$h_{31} = \frac{d_{31}}{\epsilon_3^x}(c_{11}^D + c_{12}^D) + \frac{d_{33}}{\epsilon_3} - c_{13}^D \tag{3.6}$$

The tensor relationships are quite complex, and the strain:field interactions are related through the h coefficients according to

$$h_{33} = 2\frac{d_{31}}{\epsilon_3^x}c_{13}^D + \frac{d_{33}}{\epsilon_3^x}c_{33}^D \tag{3.7}$$

$$h_{15} = \frac{d_{15}}{\epsilon_3^x} \cdot c_{44}^D \tag{3.8}$$

where the coefficients c are given by

$$c_{11} = \frac{s_{11}s_{33} - s_{13}^2}{f}$$

$$c_{12} = -\frac{(s_{12}s_{33} - s_{13}^2)}{f}$$

$$c_{13} = -\frac{s_{13}(s_{11} - s_{12})}{f}$$

$$c_{33} = \frac{s_{11}^2 - s_{12}}{f}$$

$$c_{44} = \frac{1}{s_{44}} \tag{3.9}$$

where

$$f = (s_{11} - s_{12})(s_{33}(s_{11} + s_{12}) - 2s_{13}^2) \tag{3.10}$$

These relationships together with the parameters in Table 3.5 enable us to calculate the strain:field relationships for the principal materials. These are shown in Table 3.6. For comparison the *force* field relationships are also shown in Table 3.6. These are obtained by taking the strain:field relationships and multiplying by the unconstrained compliance of the material in the direction in which the strain is applied. It is interesting to note that the strain relationships for the three principal materials PZTA, PZTB, and PVDF are very similar, but the compliance of PVDF makes it an excellent *force* transducer. Whether or not the material to be measured exerts a force or a strain upon the transducer depends upon the relative mechanical impedances of the transducer and the material to be monitored. Consequently for the case where a

Table 3.6
Typical Strain:Field and Stress:Field Relationships for Piezoelectric Materials

Material	*Field per unit strain [units Vm^{-1}]*		*Field per unit stress [units VmN^{-1}]*	
	h_{31}	h_{33}	f_{31}	f_{33}
PZT A	-3.4×10^8	0.013	2.0×10^9	0.19
PZT B	-8.2×10^8	0.002	5.5×10^9	0.04
$LiNbO_3$*	4.8×10^8	−0.002	-1.1×10^{11}	−0.9
PVDF (β)	4.0×10^{10}	0.57	-1.6×10^{11}	6.8

*Single crystal
Note. These figures indicate relative orders of magnitude; the values presented are for guidance and depend on, for example, material preparation and mechanical loading.

direct strain is to be measured (Figure 3.14(a)) there is little to choose in sensitivity terms between the PVDF and the piezoceramic material. However, if the material is to be used as a force detector—for example, in an electronic key board (Figure 3.14(b))—then the PVDF material not only gives a great advantage in terms of inherent force sensitivity but also provides flexibility and fabrication, which is impossible with ceramics.

Piezoelectric polymers can be configured as integrating [8] sensors by applying appropriate electrode structures on a sheet of material. Consequently, PVDF transducers can, for example, be made mode selective when symmetrically distributed about a mode vibration field for null sensitivity or asymmetrically distributed for maximum sensitivity or be made more (or less) sensitive at a particular spot by increasing (decreasing) the width of the contact (see Figure 3.15). While this capability is in

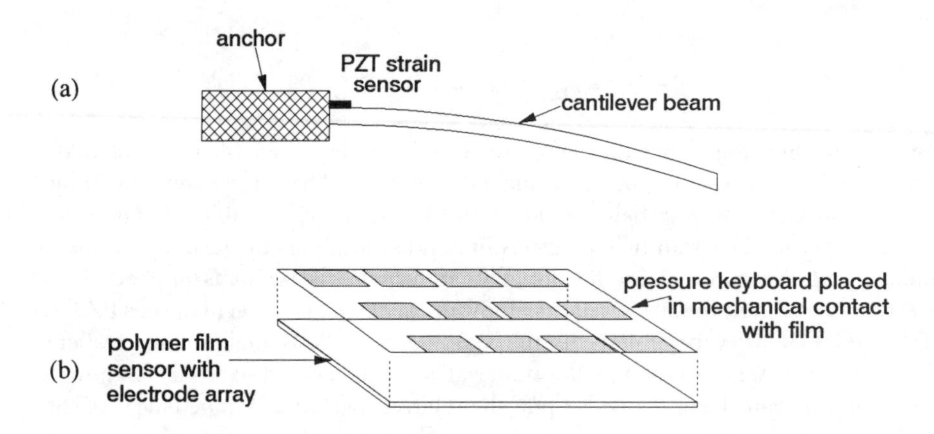

Figure 3.14 Piezoceramic as (a) a strain sensor and (b) polymer as a pressure sensor array.

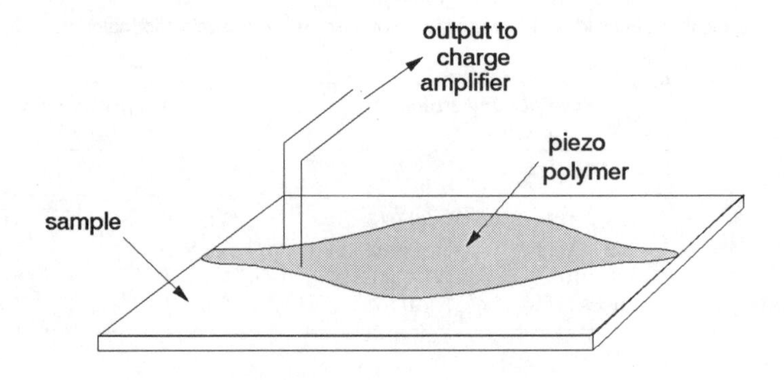

Figure 3.15 A centrally weighted integrating sensor using piezoelectric polymer film.

principle also obtainable using piezoceramics, in practice the lack of availability of large sheet materials precludes the possibility.

Strain sensing using piezoelectric materials is a well-developed art, and there is an enormous literature on the subject. This brief section has merely scratched the surface. However, even from this brief account we can begin to appreciate the subtlety and flexibility offered by the various combinations of poling and strain application axes that can be furthered enhanced, for example, through the ingenious exploitation of mechanical impedance-matching concepts that will enable the separation of strain and force fields. Yet further flexibility is permitted through the use of piezoelectric composite materials, which, in effect, are simply there to capitalize upon these inherent anisotropic interactions by incorporating appropriate impedance matching.

The potential for distributed measurements using piezoelectric materials is very limited, and no matter how they are implemented there is an implicit need for a wiring harness to each and every element within the distributed array. For large structures this can become extremely complex. In situations where integrated sensing is advantageous, sheet polymer with an appropriate electrode pattern can be configured to act in this fashion.

The aim of this section has been to give an insight into the properties and applications of piezoceramic materials. The bibliography cites references to some of the more important contributions to the extensive literature on the subject.

3.4.2 Inductively Read Transducers—The LVDT

The LVDT is included here as a much-loved, much-used measurement device that has been in the catalogs for decades. The basic idea is very simple (Figure 3.16), and LVDTs are now a regularly available, very well engineered product capable of very precise *displacement* measurement.

While they do have obvious applications limitations and in particular can only perform point measurements and are relatively cumbersome physically, they have been used extensively in monitoring civil engineering structures and could be construed as contributing to the first generation of "smartness" in such structures.

It is tempting to argue that LVDTs will be superseded by other technologies since they are complex to build and restricted in their use. However, the measurement community is accustomed to using these devices: they can be made in a very temperature-stable engineered form, and they have a proven reliability and accuracy record. There are some situations in which the LVDT is *exactly* the way in which a measurement should be implemented [9].

3.4.3 Fiber Optic Sensing Techniques

To many, fiber optic sensing techniques appear as the most promising sensing technology available for smart structures use [10]. Indeed, whole texts have been devoted to the subject [11].

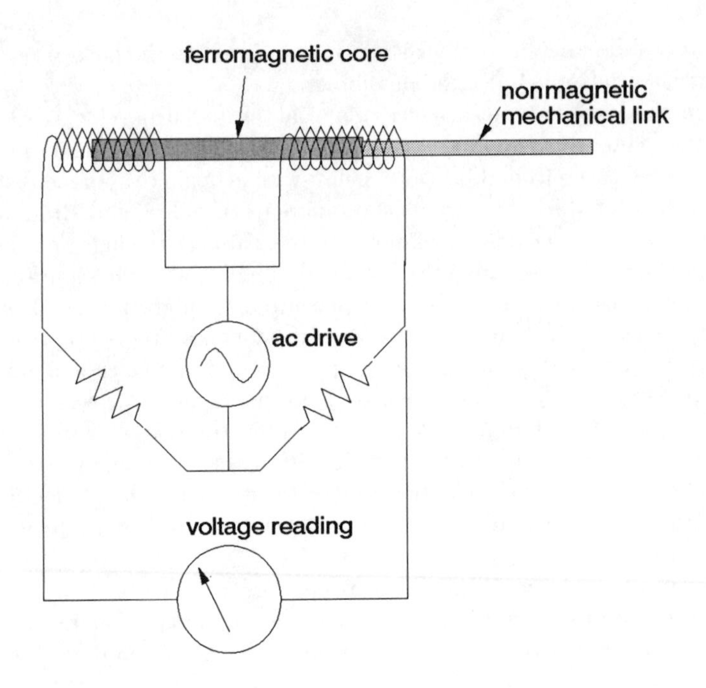

Figure 3.16 The elements of LVDT displacement sensing. Displacement is read as an imbalance in the detection bridge.

Historically optical fibers have been applied to monitoring physical properties in structural applications, but more recently the techniques have also been expanded to encompass important chemical measurands. In addition, in by far the majority of structural monitoring applications, the fiber itself is used as both the transducer and the communications medium. In effect, the fiber comprises an integration of the entire sensor system hardware within a single cable. By appropriately designing the cable/fiber combination, systems may be designed to interface with most important structural monitoring measurands.

The other important feature of optical fibers is their *materials compatibility*. In the majority of structural monitoring applications the immediate environment is relatively harsh due to either corrosion and corrosion products, the possibility for electrolytic action, the high pH values experienced in concrete, or simply the ingress of moisture accompanied by uncontrollable additives (for example, salts, solvents, and cleansing fluids). The vast majority of metal-based communication and sensing structures will eventually corrode under almost any of these conditions. However, the silica optical fiber is inherently stable provided that the final hermetic coating on the fiber does its job.

Another benefit from optical fibers that is not available with other media is their intrinsic immunity to electromagnetic interference. For comprehensive monitoring systems, especially those in applications such as airframes, electric trains, and civil structures such as building and bridges, this immunity to electrical interference effects is an extremely powerful advantage.

Another extremely attractive feature of fiber optic sensing is the essentially lossfree transmission medium. This is particularly important in relatively large structures (say greater than a few meters) since it implies that no remote power supplies are needed at any of the sensing points in structures up to tens of kilometers in dimensions.

The remainder of this section will examine the basic transduction systems available for optical fiber measurements applicable to structural monitoring and continue to discuss the basics of the many multiplexing possibilities. We shall continue to describe some examples of fiber optic sensing systems with a view to assessing the real possibilities for particular approaches to both chemical (in Section 3.5) and physical measurements.

3.4.3.1 Sensors and Sensor Systems Using Optical Fibers

The overall topic of optical fiber sensors is extremely diverse, and as for piezoelectrics, many texts have been devoted to the subject [12, 13]. In this present context we must attempt, at the risk of omitting something important, to distill the topic down to its essential features for structural monitoring.

Only temperature, strain (or length), and (usually) corrosion-related chemical measurements are involved. These measurands may be imposed upon an optical carrier through either linear or nonlinear interactions. Here a *linear* interaction is one that is detected at the same optical frequency as the probe light, while a *nonlinear* optical interaction is one that is detected as a different optical frequency from the probe light. One of the unique features of optical fiber measuring systems is that an extremely rich variety of transduction techniques are available. Used with care, this can give unrivaled flexibility since, for example, in principle both strain and temperature may be measured using a single optical fiber but different probe techniques.

We should also restrict most of our deliberations on fiber sensors for structural monitoring to systems that involve only the fiber as the transduction and communications mediums. Under these constraints the principal technological options are those summarized in Table 3.7.

This section is focused upon physical parameter measurement, namely strain and displacements. However, it is particularly important with fiber optics to recognize that there is often the possibility for interference with an apparent strain reading from other sources often induced by temperature changes. Table 3.7 considers the three (temperature, stress, strain) important measurand domains. These cross sensitivities affect other sensor systems, but the impact for optical fiber systems is frequently much more profound.

Table 3.7
Optical Fiber Techniques for Structural Sensing

Optical Technique		*Measurand*		
Linearity	*Mechanism*	*Temperature*	*Strain*	*Chemical*
Linear	Delay Modulation	Optical phase/differential match phase, subcarrier phase	As temperature	As temperature but via chemical interface
	Intensity Modulation	Microbend via thermal transformer	Microbend via strain transformer	Microbend via chemical transformer
Nonlinear	Raman	Stokes:anti-Stokes ratio within fiber	Stokes, via transforming material outwith fiber	Stokes, via probe into material
	Brillouin	Backscatter frequency, in fiber	Backscatter frequency, in fiber	N/A except via chemical-to-mechanical converter
	Kerr	Used to generate a probe signal	T/ϵ variation of Kerr coefficients appear low	N/A
	Fluorescence	Via intermediate material fluorescence decay	N/A	Via intermediate chemistry affecting fluorescence

Table 3.7 makes a number of important points:

- Linear modulation systems, while inherently simpler and certainly more efficient in terms of achievable optical powers at the receiver, are subject to optical reflection phenomena that must be taken into account during the design of the sensor system to ensure that they do not swamp the measurand signal. Consequently, it is often difficult to use linear modulation in reflection-based systems unless these reflections from irrelevant parts of the optical path can be readily gated out.
- Nonlinear systems involve a generation of signals involving light propagating down the fiber being absorbed and re-emitted. The relatively small absorption cross section plus the fact that the re-radiated emission is almost always isotropic imply very low efficiency (typically less than 1%). Thus the high power is needed not only to stimulate the process within the material but also to ensure that a sufficient return signal is generated. However, the fact that the return signal is at a *different* optical frequency means that it is usually straightforward to eliminate spurious responses.
- The essential features of nonlinear and linear transduction processes are shown in Figure 3.17.

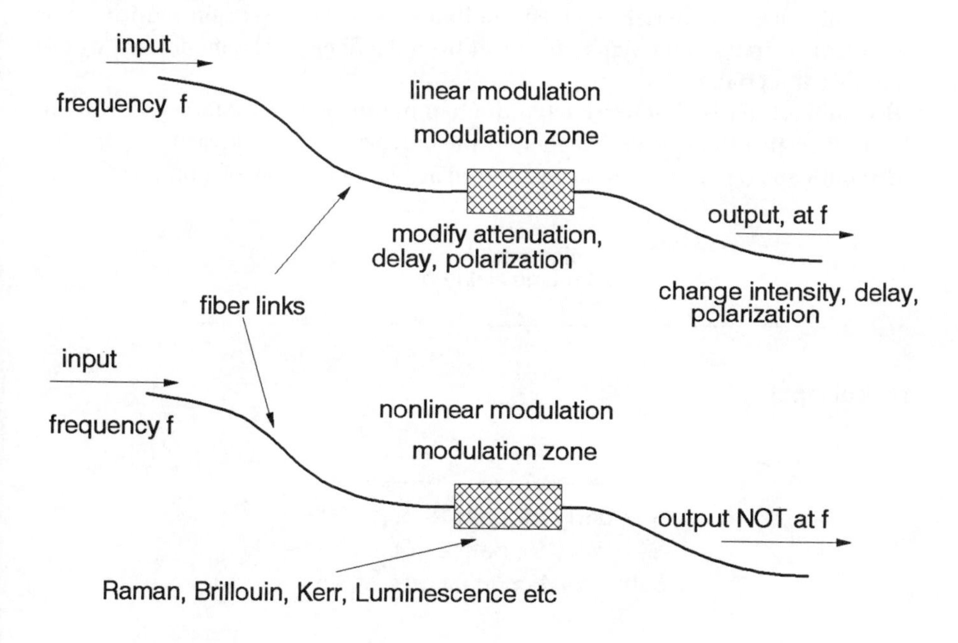

Figure 3.17 Linear and nonlinear optical processes for measurement.

Within the subdivisions of linear and nonlinear techniques we may make the following observations:

- Delay modulation may be either *directly measured* or measured through the *differential* between two paths in the fiber. For the former, the 1° temperature change is broadly equivalent to the 10 μstrain change. For the latter, the temperature:strain ratio can be modified depending upon the modes selected and the structure of the fiber. Delay measurements are essentially independent of the chemical environment unless some transformer is introduced (Figure 3.18).
- Intensity modulation (Figure 3.19) is only introduced as a distributed parameter through microbend phenomena. The interesting point is that the microbend transformer may be configured in a rich variety of ways and optimized to monitor temperature, temperature transitions, strain, or the presence or absence of chemical species.

Similar comments may be made about the nonlinear interactions:

- Raman scatter has the advantage that it is a characteristic of most materials that varies only with temperature. In the case of silica, therefore, it can be used to very good effect as a temperature probe [14]. To access strain requires carefully chosen material in which the Raman spectrum is strain sensitive. For strain measurement the signal must exit from the fiber and be modulated within another material.
- Brillouin scatter is closely related to Raman but involves an interaction with an acoustic rather than an optical phonon and only produces backward propagating Brillouin scatter. Since the acoustic velocity is a function of both strain and

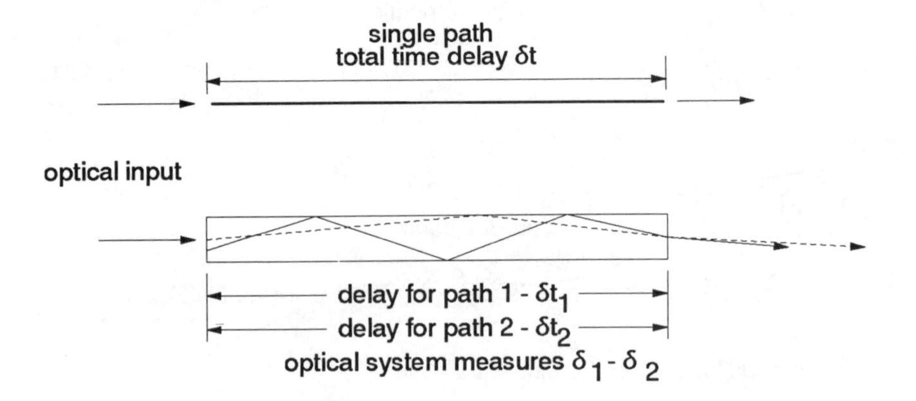

Figure 3.18 Integral and differential delay measurements for path length measurements.

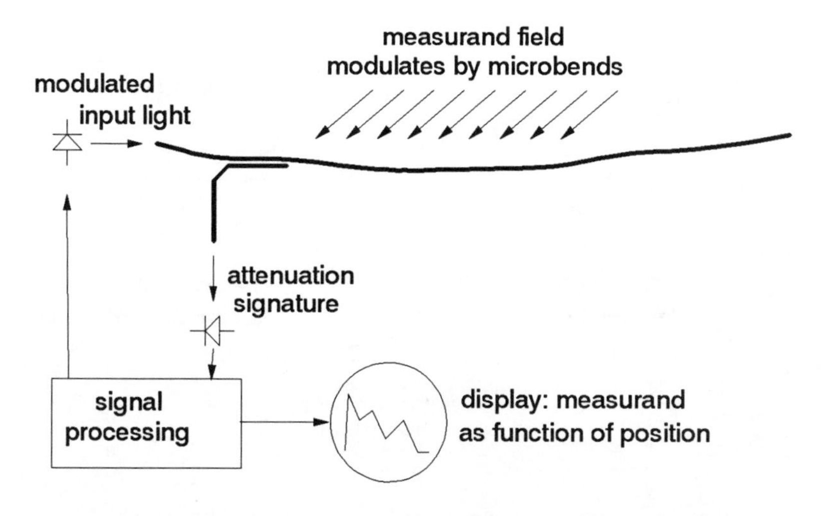

Figure 3.19 Measured modulated microbends for backscatter modulation giving distributed measurements. Similar comments may be made about the nonlinear interactions.

temperature, the backscatter frequency is similarly a function of strain and temperature [15].

- The Kerr effect is an inherent nonlinearity in the refractive index of the material and will *either* stimulate second harmonic generation *or* produce mixing of two optical signals within the fiber to produce sum and difference frequencies. Neither has, as yet, been used seriously in instrumentation, but preliminary experimental results have demonstrated that the value of the Kerr coefficient *is* strain sensitive but *not* very dependent on temperature [16].
- Fluorescence in its many manifestations is essentially a chemical phenomenon, though the fluorescence decay times of many materials are often functions of temperature. There are few, if any, means whereby a strain field may be turned into fluorescence variations, but temperature probes have been fabricated using the thermally induced variation in the decay constants.

Figure 3.20 shows the principal features of these nonlinear processes. All may in principle be used as the basis for distributing measuring systems. It is this capacity for distributed monitoring that is unique to optical fiber sensor technology and is exceptionally useful in structural instrumentation since it totally obviates the need for complex wiring harnesses within the structure. Additionally it removes the need for electrical, hydraulic, or other sources of power at each sensing point. Similarly individual fiber optic sensors may be configured as totally passive multiplexed networks with sensor interrogation and decoding located at a central point.

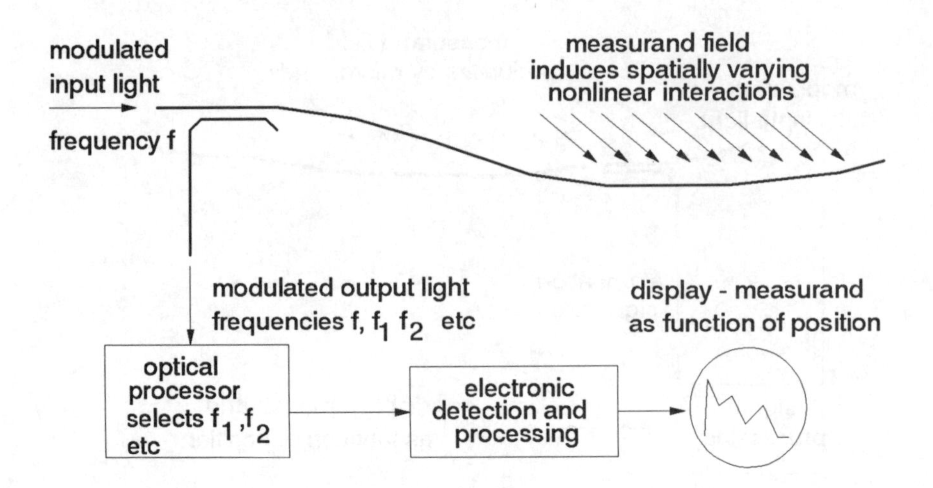

Figure 3.20 Use of nonlinear phenomena for distributed sensing—general features.

In summary the principal features of fiber optic sensors and systems for structural monitoring are the following:

- The rich variety of linear and nonlinear transduction techniques may be interfaced to most if not all of the pertinent measurands often on the same fiber.
- The sensing system is chemically inert, mechanically extremely tolerant (very high elastic limit), and totally immune to the impact of electromagnetic interference.
- The sensors themselves are immediately compatible with operation in distributed, quasi-distributed, or multiplexed modes with no requirement for electrical or other sources of power within the sensor area.

3.4.3.2 Fiber Optic Strain Sensing for Structural Monitoring

The principal techniques that have emerged for this application are:

- Optical delay modulation systems relying on phase, differential phase, or subcarrier phase and configured usually as quasi-distributed elements comprising either suitably delineated fiber sections or Bragg gratings or similar at predetermined points;
- Microbend modulation in devices using suitable transformers configured in a distributed format;
- Brillouin scatter systems configured as distributed sensors.

Some preliminary measurements have also been made using the Kerr effect in a pump:probe distributed configuration and Raman scatter from organic sensor materials in a point transducer configuration.

Optical delay modulation is the most direct of these. However, all direct optical phase modulation systems are influenced by both temperature and strain variations. In all cases a temperature change of 1°C produces approximately the same phase change as strain change of 10 strain. Even a modest operating temperature range of 40°C imposes an automatic uncertainty of 400 μstrain in the perceived value, so some form of compensation is essential to achieve the desired strain resolutions, typically in the few tens of microstrain region.

The most usual configuration for a phase delay strain sensor is shown conceptually in Figure 3.21. The fiber is mounted at the points A and B within the structure, and the optical system measures the time taken for the optical signal to travel between these two points. This is then related to the strain experienced by the structure. However, we should note that normal changes in temperature will also produce an apparent strain on the sensing fiber since the linear coefficient of expansion for silica (of the order of 0.5×10^{-6}/°C) is significantly less than that for most structural materials where figures of the order of 10 parts/million/°C are typical. Indeed the mounting conditions impose approximately the same differential phase with temperature as the inherent differential phase within the sensor itself. Thus, normal temperature excursions of almost any structure will impose a strain upon the optical fiber. To cope with temperature excursions below normal ambient, the fiber sensor must be installed under tension within the structure or must be suitably constrained by the structure in order to impose a compressive strain. Consequently disentangling the temperature and strain measurement in any structural monitoring application is essential in order to convert the actual strain experienced by the structure from the strain experienced by the sensor at the measuring temperature. Usually it is only the strain that is of interest since it is rare for a structure to be compromised by unusual

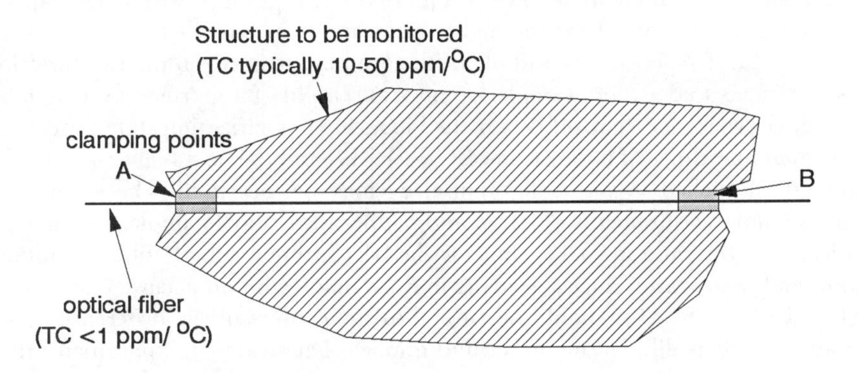

Figure 3.21 Interactions between fiber strain/distance measurement and structures.

thermal excursions. The temperature compensation technique described below—or indeed referring to any other sensor technology—must always be viewed with these differential thermal phenomena in mind.

It is evident that any thermal compensation scheme must *either* use a strain transducer that is itself totally thermally insensitive *or* measure both temperature and apparent strain on the fiber simultaneously. While the former is, in principle, attractive, the above discussion on temperature effects does imply that its usefulness will be limited to occasions where temperature excursions are relatively low—where thermal compensation is not necessary anyway. If we need to measure both temperature and strain and it is obvious that two delay measurements are necessary, these can be expressed in the form

$$\begin{bmatrix} \varphi_1 \\ \varphi_2 \end{bmatrix} = \begin{bmatrix} \partial\varphi_1/\partial T & \partial\varphi_1/\partial\epsilon \\ \partial\varphi_2/\partial T & \partial\varphi_2/\partial\epsilon \end{bmatrix} \begin{bmatrix} \Delta T \\ \Delta\epsilon \end{bmatrix} \tag{3.11}$$

If the 2×2 matrix on the right-hand side of this equation can be inverted, then ΔT and $\Delta\epsilon$ may be resolved. The condition for inversion reduces to

$$\frac{(\partial\varphi_1/\partial T)/(\partial\varphi_1/\partial\epsilon)}{(\partial\varphi_1/\partial T)/(\partial\varphi_2/\partial\epsilon e)} \neq 1 \tag{3.12}$$

The further the ratio of (3.12) differs from unity, the more accurate the inversion process from a given achievable measurement accuracy of φ_1 and φ_2; the matrix must be well conditioned.

These basic observations have lead to the use of a number of *differential* delay techniques in structural monitoring and, in particular, to the use of elliptical-cored optical fiber as the sensing medium exploiting the unusual birefringence and modal dispersion properties of this particular medium. The basic principles of one particular measurement technique are shown in Figure 3.22. Light at a wavelength λ_1, which is longer than the cut-off for the higher order mode in this fiber, is launched at 45 deg to the birefringence axis and the differential phase between the two birefringent modes is measured as, for example, φ_1 in (3.11). Light of a shorter wavelength below the higher order mode cut-off is launched in a single polarization state in such a way that equal intensities of both the lowest order LP_{01} and the next higher so-called LP_{11} mode are launched. Again a differential φ_2 may be measured. The accuracy with which actual measurements of strain and temperature may be implemented depends critically on the ratio in (3.12), but for the best fitting available fiber simultaneous strain and temperature measurements may be made with an accuracy of ± 1°C and ± 10 microstrains. This should be viewed in the context of the differential thermal expansion effects alluded to earlier and implies that strain—the parameter that we really need—may be evaluated over a range of temperatures to a few tens of microstrains using this technique.

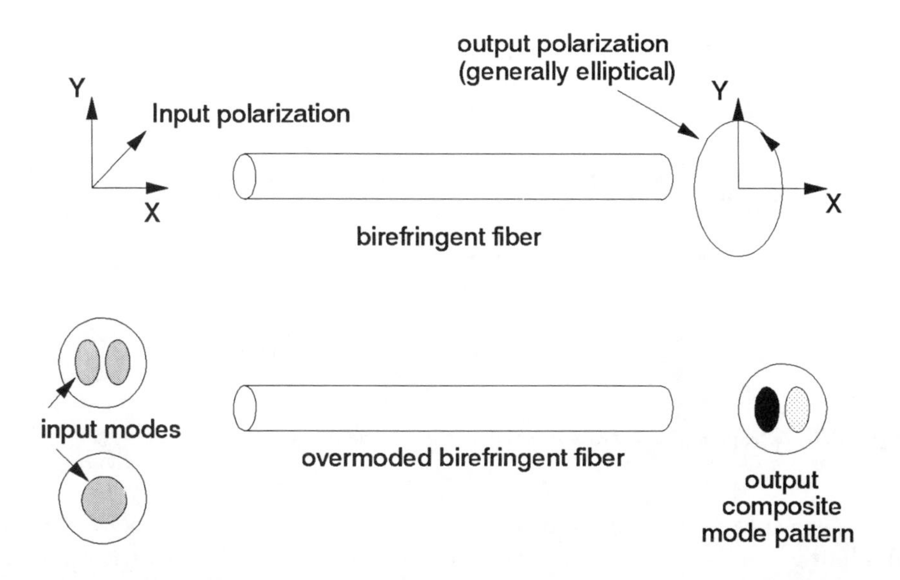

Figure 3.22 Basic concepts of polarization and mode:mode sensing systems for combined temperature and strain measurements.

It is, of course, possible to ring the changes in the basic configuration shown in Figure 3.22. In particular, it is useful to operate at the one wavelength and perhaps examine φ_1 and φ_2 as LP_{01}/LP_{11} interferences in two different polarization states or indeed examine the interference between the two LP_{11} modes as one of the variables. More recent research on these interactions has indeed pointed the way toward more readily implemented approaches that might lend themselves to an all-waveguide optical launch and a demodulation system. The system shown in Figure 3.22 requires quite complex launch-and-receive optics in order to implement the measurements.

A variation on this basic theme is to examine a combination of delay and dispersion in a single-mode fiber: The latter is much more sensitive to temperature than strain compared to the former [10:1 to 100:1 approximately]. Dispersion measurements can be made using a broadband source, selecting three bands within this broadband, and measuring the differential delays between adjacent pairs of bands [17].

The remaining techniques that have been investigated only lend themselves in their current form to a single delay measurement, so temperature compensation is either ignored or extracted from the separate temperature measurement.

Of the direct delay measurement techniques probably the simplest involves the use of radio or microwave frequency subcarrier modulation (Figure 3.23). The system may be configured either as a phase measurement device or with the fiber acting as the feedback component in a ring round oscillator. Both techniques have been used

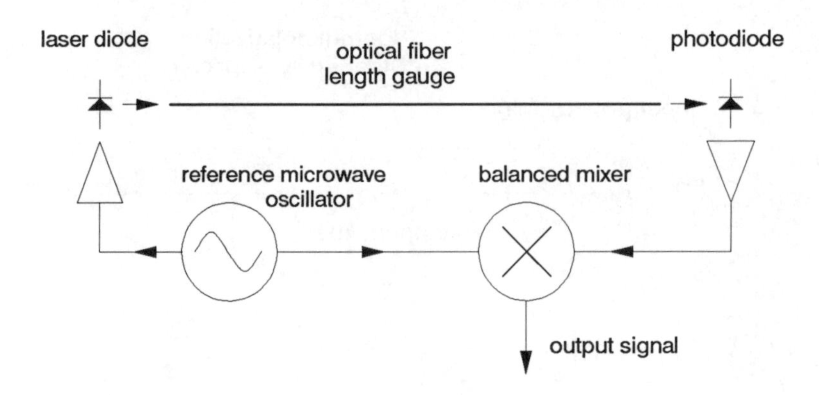

Figure 3.23 Basic principles of microwave phase optical length gauge.

with some success, though the phase measurement system has possibly the greater flexibility. Similar systems have been used as range finders for surveying purposes for some considerable time. The accuracy of phase measurement devices depends upon the phase meter, its stability, and the subcarrier modulation frequency. Typically with subcarrier frequencies in the region of 1-GHz measurement resolutions of somewhat better than 100 μm are practical with a very simple phase detection circuit. There is obviously an ambiguity in the absolute measurement of length corresponding to—depending on the phase bridge—one half-wavelength of the subcarrier frequency in the fiber. At 1 GHz this corresponds to a 10-cm ambiguity. This ambiguity can, however, be resolved by using measurements with subcarriers at considerably lower frequencies or, of course, by using appropriate prior knowledge of the expected length ranges of the sensing fiber. The accuracy of the measurement is determined finally by the stability of the reference oscillator. However, the direct phase measurement can resolve the limit one part in 10^4 of the wavelength. The achievable resolution therefore behaves in the manner shown in Figure 3.24, where we have assumed that the oscillator stability can be guaranteed to 10 parts/million. In essence the *strain* resolution varies from 100 μstrain for a gauge length of one wavelength to 10 μstrains for gauge length exceeding 10 wavelengths. The achievable resolution can be enhanced for long gauges measured in wavelengths, provided that the oscillator stability can be improved. Clearly the usual comments on temperature must also apply. In particular, to achieve 10-μstrains resolution implies a knowledge of temperature to within 1°C.

A very closely related technique involves the use of optical time domain reflectometry as the measuring system (Figure 3.25). The achievable resolution here is directly dependent upon the length of the pulse used to probe the network. State-of-the-art (but extremely expensive) optical time domain reflectometers are capable of producing picosecond pulses with measurement accuracies of the order of

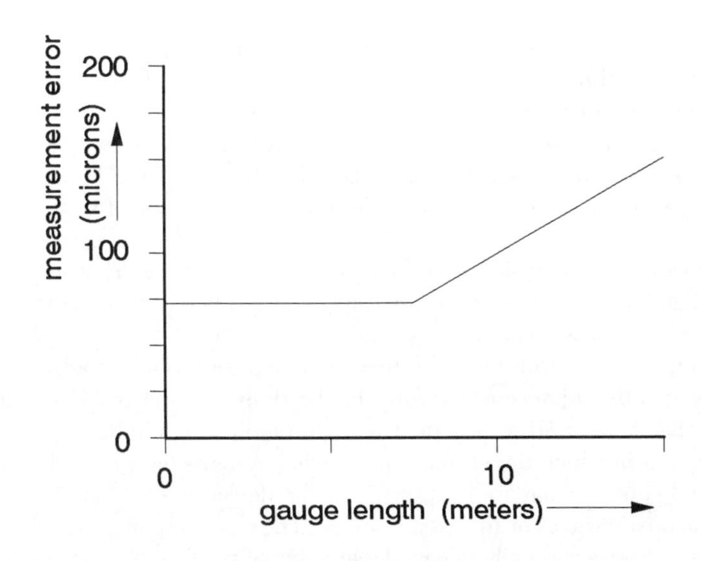

Figure 3.24 Measurement error versus gauge length for 1-GHz subcarrier frequency and 0.1° measurement error.

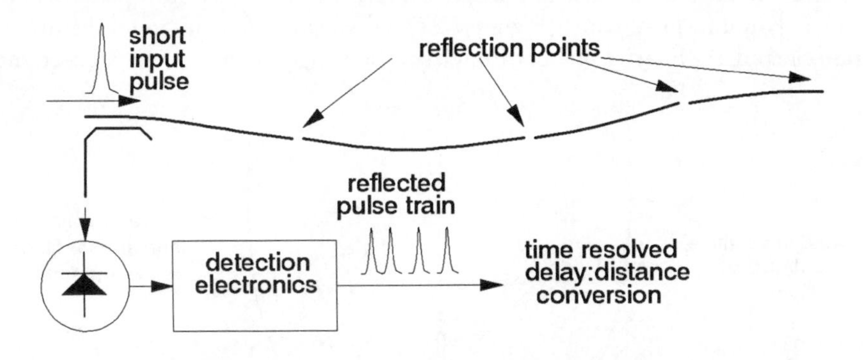

Figure 3.25 OTDR-based distance measurements: resolutions better than 100 μm are feasible.

100 μm. However, in contrast to the radio frequency subcarrier technique this is an extremely expensive probe.

Direct measurements of optical phase can also be useful indicators. However, the fact that 1 μstrain over 1m corresponds to a changing of phase by 10 rads at typical optical wavelengths implies that such measurements are only usable over extremely short gauge lengths. Consequently, direct measurements of optical phase are not usually appropriate for the gauge lengths of the order of meters that typify structural monitoring applications. However, for point systems where localized strain needs to be monitored over gauge lengths of millimeters, direct optical phase measurement is a very useful technique. Figure 3.26 shows a Fabry Perot rosette strain gauge that may be used to extract all the appropriate components of the strain tensor at a particular point. This strain gauge operates by interfering between beams that reflect from each end of the measuring fiber. Despite the use of the Fabry Perot term, it may be regarded as more close in operation to a twin-beam Fizeau interferometer [18].

The so-called extrinsic Fabry Perot interferometer (EFPI) is similar in concept but uses a gap about one hundred microns in length between two single-mode fibers as the resonator. In this case its response is determined entirely by the material that holds the input and output fibers in their relative positions and not by the fiber itself. This gives greatly enhanced flexibility in the design of the response function. The basic idea has been used widely in structural monitoring applications.

One area in which direct phase measuring systems have considerable potential (and indeed considerable application) is in the detection of dynamic variations, for example, acoustic waves or ultrasonic waves. The relatively small displacements that typify these waves are readily detected using one of the interferometer configurations shown in Figure 3.27. In designing these interferometers the $\cos^2$ response that characterizes all of them must be taken into account. If the phase deviations are relatively small (and usually they are), then the interferometer must be biased at one of the maximum slope points (Figure 3.27). Often static changes in temperature and strain can pull the interferometer through several fringes so that an ingenious, dynamic

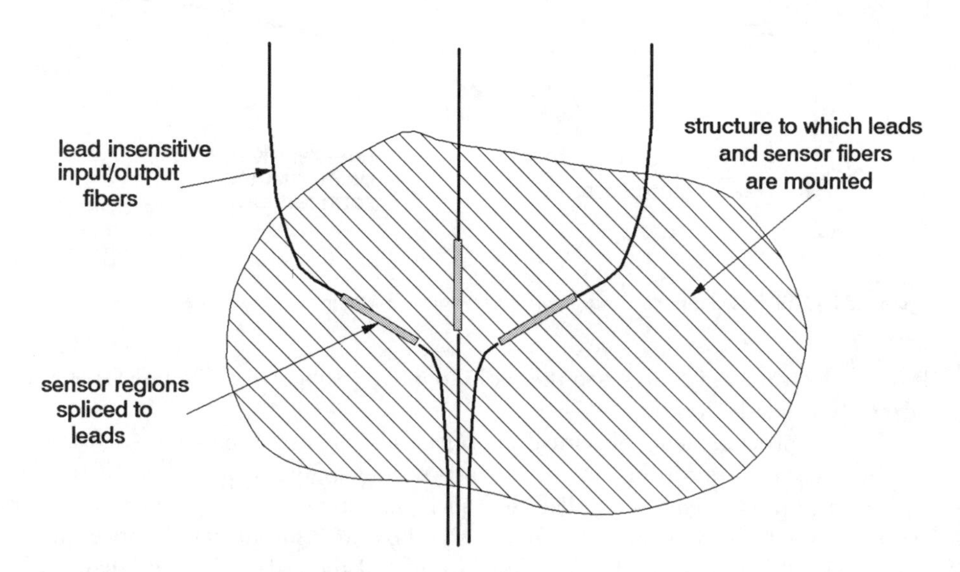

Figure 3.26 Optical fiber strain rosette designed to access all strain directions (after Valis, UTIAS).

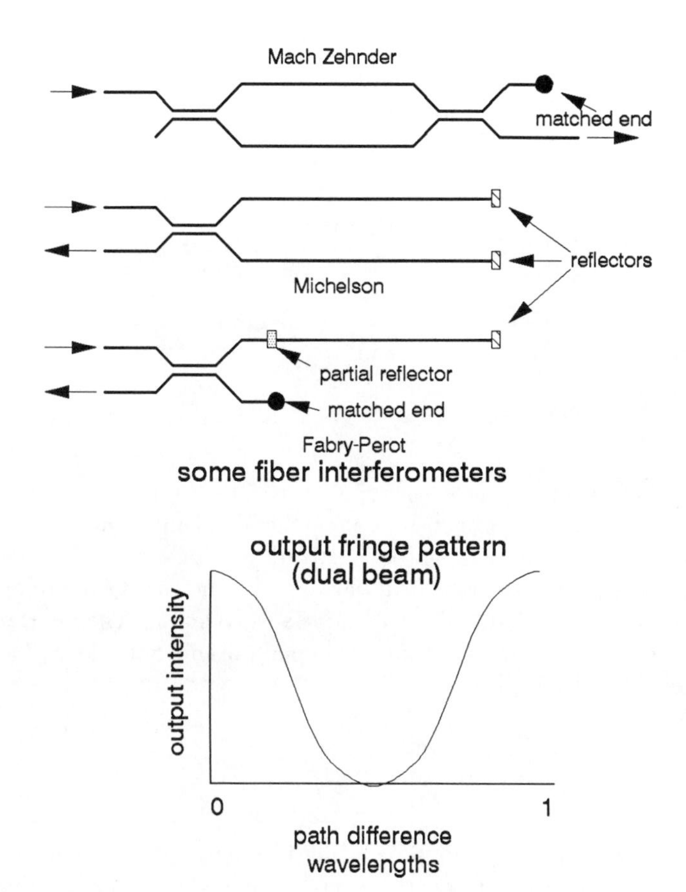

Figure 3.27 Dual-beam fiber interferometers for structural measurement.

demodulation technique or the use of a three-phase detection system using a 3×3 coupler is essential. It is beyond the scope of this text to go into the details of these arrangements but much has been written elsewhere on the topic [12].

The so-called Bragg grating capability of optical fiber technology has introduced another degree of freedom into sensor design. The name derives from the father-and-son team who pioneered x-ray crystallography and discovered that for particular orientations of crystals with respect to a quasi-monochromatic x-ray source virtually all of the radiation could be deflected into a particular direction that was determined by the separation of the crystal planes within the structure under examination. In optics Bragg diffraction refers to the ability of thick phase diffraction gratings to have exactly the same effect.

In the optical fiber context the Bragg grating is formed by introducing changes in the core refractive index along the length of the fiber (Figure 3.28). This phase

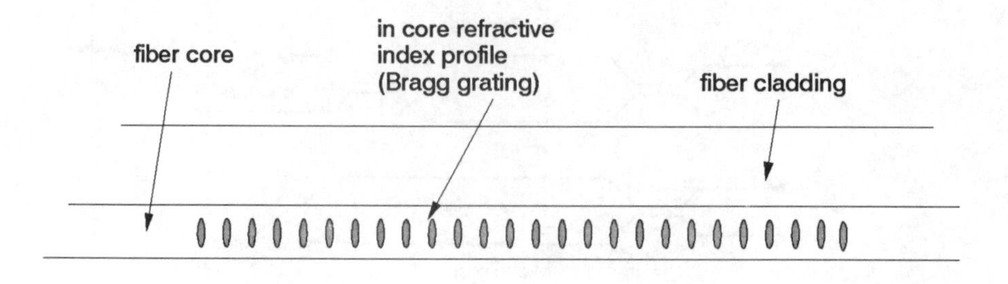

Figure 3.28 Basics of in-fiber Bragg grating geometry.

grating may be introduced in a variety of ways; for example, the standing wave produced by counterpropagating beams along the fiber length may induce the grating through the Kerr effect. However, most usually the gratings are fabricated by inducing a still incompletely understood optical damage into the core of the fiber using high power density ultraviolet light. The technique is very simple. The interfering beams, typically from an excimer laser, are focused upon the core and the wavelength of the grating within the core is given by

$$\lambda_{\text{core}} = \frac{\lambda_{\text{source}}}{2 \sin \phi_L} \tag{3.13}$$

The behavior of these gratings, particularly in terms of their frequency response, is somewhat different in detail from the behavior of the classic thick grating. However, we can make some useful approximations that form very reasonable guidelines on how these gratings behave.

Suppose the grating comprises N wavelengths along the core of the fiber with a peak refractive index variation of Δn. Such a grating will provide a reflection with a peak value occurring at

$$\lambda_{\max} = 2\lambda_{\text{core}} \tag{3.14}$$

and the transmission will have a notch at this same wavelength.

The fractional power reflected at $\lambda_{\max}$ will be of the order $(N\Delta n)^2$, and the bandwidth *assuming that the grating itself is perfectly periodic* will be of the order of N^{-1}. Of course, once the total reflected power has achieved its maximum value, the remainder of the grating in effect plays no real part in its operation at this wavelength, so this reflection bandwidth will saturate at a greater value than N^{-1} if these conditions are met.

Several interesting results have been achieved with Bragg gratings both as the sensing elements themselves and as the coupling elements in quasi-distributed sensor arrays. When used as the sensor element, the normal strain and temperature interference considerations apply. The read-out system is usually centered upon measuring the reflection wavelength, and clearly this will scan through a range equivalent to the maximum strain imposed upon the sensor. Depending upon the material, this can range from 0.5% to 5%. The fundamental resolution that can be achieved depends upon the detector's ability to locate the center of the reflected spectrum (or an equivalent operation). For the simplest type of detector, this resolution will be of the order of N^{-1} (that is, the bandwidth), though with improved systems, a centroid location from 1% to 10% of the bandwidth is achievable. The *noise-limited* center location of the spectrum can be parts in 10^6 or better, depending on the optical power within (or excluded by) the filtering action (see Figure 3.29). This limit has been achieved in other applications, for example the ring laser gyroscope, but not as yet with Bragg grating detectors. Electronically scanned spectrometers could make a great impact here. Typical maximum values for N are in the order of 5,000, so in principle strain resolutions in the region of 10 μstrain (or, alternatively, sensitivities to temperature changes of 1°C) may be achieved in quasi-static measurements. However, the question of a stable, simple wavelength reference remains open, though there has been some interesting activity based on calibrated FTIR spectrometers [19].

There are other considerations that apply to the use of Bragg gratings. The need for grating period uniformity along the length is paramount, and any jitter results in both a reduction in the peak-reflected power level and an increase in the

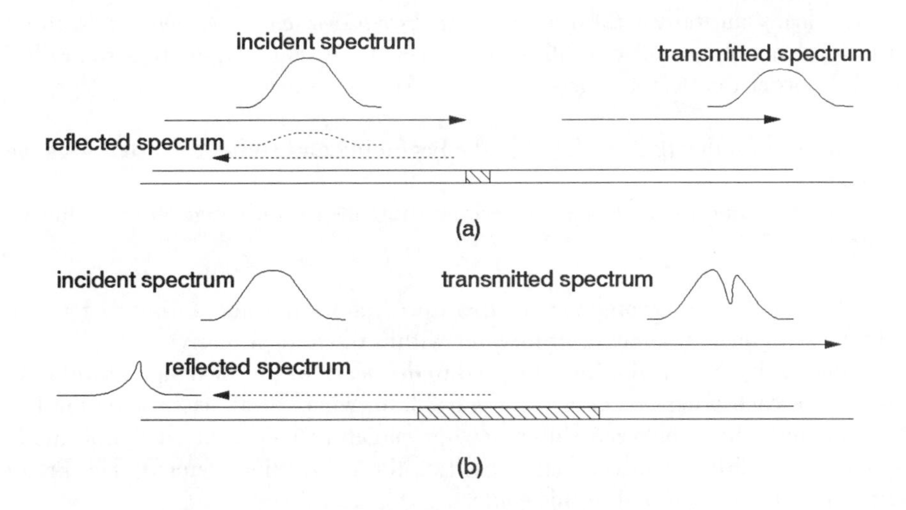

Figure 3.29 Functions performed by in-fiber Bragg gratings: (a) short broadband reflector/mode coupler; (b) long narrowband wavelength selective coupler.

achievable bandwidth of the filter. Additionally, the process by which these gratings are formed is far from understood, and the debate continues over issues such as long-term stability, the impact of temperature cycling, and the susceptibility to bleaching during use. That said, a considerable body of empirical data indicates that these factors can indeed be accommodated.

There is also the very considerable potential benefit that these gratings may be realized during the fiber-pulling process by attaching a pulsed excimer laser together with the appropriate optics onto the rig. Preliminary experiments in this direction have given encouraging results.

In summary, the Bragg grating offers the prospect of a readily fabricated point sensor that can be incorporated within the fiber structure during the fabrication process. This sensor—given an adequate wavelength-discriminating receiver—should be capable of resolving a few tens of microstrains but requires careful mounting procedures to ensure that the grating periodicity remains uniform and the recognition within the detection electronics that the grating index depth may well deteriorate with age. For sensors gratings with N of 5,000 or thereabouts are required.

These gratings may also be used to good effect as mode-coupling structures either to reflect a small amount of power or to couple a small amount of power from one mode to another in a low-mode fiber sensor such as that shown in Figure 3.22. In this context N of the order of 10 provides coupling at the level of about −20 dB to −30 dB over a bandwidth that is now of the order of 10%. Furthermore, the value of this coupling coefficient is largely independent of temperature or strain applied to the fiber but of course may deteriorate with aging depending upon the characteristics of the grating.

This leads naturally to the topic of *quasi-distributed sensor systems*, and without this facility the applicability of all of the direct-delay sensor systems is somewhat limited. In order to effect a quasi-distributed system we need:

- A means of indicating on the fiber the beginning and end of each sensor gauge length;
- A means within the receiver of discriminating one sensor gauge length from the next.

Additionally, for a combined temperature- and strain-measuring system we need a measurement channel discriminator within the optical receiver.

Some of the options that may be used to delineate strain (or temperature) only sensors from each other are shown in Figure 3.30, where in all cases some kind of mode coupler (either between different copropagating modes or the same mode propagating in different directions) is used as the delineation element. The Bragg grating is particularly useful in this context.

The simplest quasi-distributed system is that using short pulse OTDR (Figure 3.25) as the interrogation mechanism. This enables 100-μm to 200-μm resolution

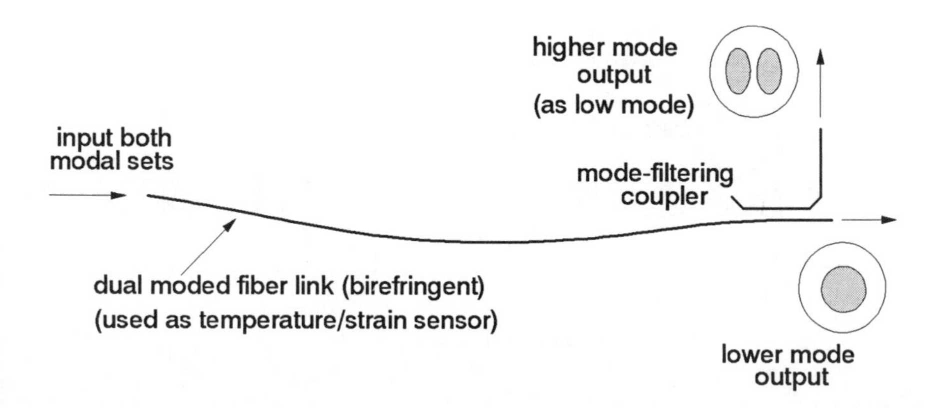

Figure 3.30 One configuration for all fiber strain/temperature monitoring. The output ports may feed to polarizers to demodulate.

of the lengths of each of the sensing elements. Similarly a multifrequency subcarrier system may be used and, provided that the operating frequencies are chosen carefully, can also provide 100-μm to 200-μm resolution of each sensor gauge length.

A multielement sensor system using mode:mode interference or birefringence may also be configured using the same approach; for example, light may be launched on a specified polarization axis of a birefringent fiber operated at wavelengths beyond the cut-off. A small amount may then be coupled to the orthogonal polarization axis at each coupling point and a white light receiver used to demodulate the output. This system has been demonstrated to be capable of resolutions of a few microstrain on gauge lengths of the order of 20 cm.

Two-channel systems capable of temperature and strain extraction may be realized by extending the same basic principle. Here light is launched on a single mode in a single polarization state, and the mode-coupling systems now serve to couple into two different types of measurement that may involve, for example, LP_{01} interference in two orthogonal polarization states (birefringence) and the interference of one of these polarization states with an LP_{11} in the same state, though other combinations may prove to be much more convenient. However, given the right type of fiber, a multiplexed temperature and strain system may be realized. Additionally it may be fabricated from all waveguiding components, assuring the potential for ruggedness and portability.

Direct optical delay measurements in optical fibers enable the realization of a wide variety of measurements. All of the measurements are to some extent affected by both temperature and strain fields, so some compensation mechanisms are essential. Given the fact that a thermal change may introduce its own spurious strain on a silica-based strain sensor, combined temperature and strain measurements are

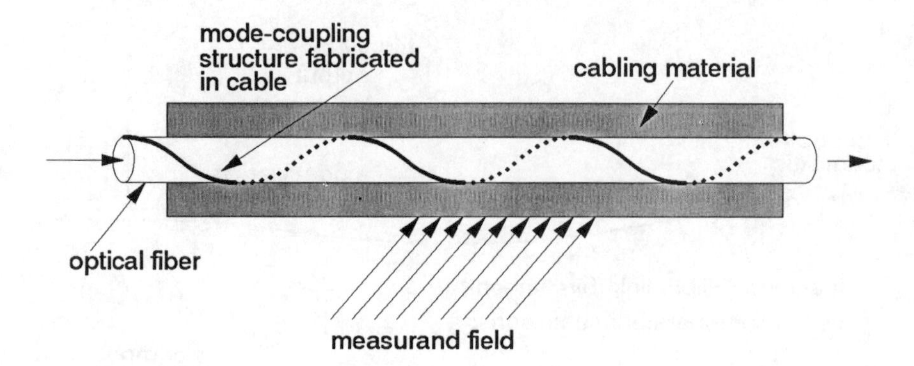

Figure 3.31 Schematic of strain-sensitive microbending cable. Deformations to the cabling material are transferred to microbending-induced losses via the coupling structure.

desirable in order to remove this effect from the measurements. However, for temperature-stable environments (defined as those in which induced temperature effects are less than the necessary strain resolution) very simple subcarrier-based measurements promise short-term practical realization. In the medium term the quasi-distributed systems offer substantial potential for dynamic measurements where thermal effects are too slow to affect the readings. There is significant longer term potential for quasi-static measurements based upon the same technique but with dual-channel interrogation. In all cases a complex sensor array may be addressed through a single optical fiber port. The projected number of elements in each array is currently felt to be up to a few tens, and present systems achieve acceptable performance with up to 10 elements in them.

3.4.3.3 Strain Sensing Using Intensity Modulation

Strain sensing-producing intensity modulation in a distributed or quasi-distributed architecture is only tractable through microbend sensors (Figure 3.31). The principles of microbend modulation are essentially to surround the fiber by a geometry that introduces a strain-dependent periodic perturbation along the path of a multimode optical fiber preferably based upon a graded index geometry. The mode distribution within such a fiber can be described through their characteristic propagation constants, and these propagation constants differ by an almost constant factor $\Delta\beta$. If the periodicity of the imposed disturbance corresponds exactly to this difference in propagation constants:

$$l_{\mathrm{p}} = \frac{2\pi}{\Delta\beta} \tag{3.15}$$

where l_p is the period of the disturbance, then a very small perturbation will cause significant coupling between these modes. Some of the coupling is to radiated modes, so the net effect is to impose a localized loss increase, the value of which depends upon the strain at a particular point in the fiber. Probing this structure with an optical time-domain reflectometer gives a backscattered power signature that depends upon the local strain. Depending upon the type of OTDR in use, the resolution in this backscattered signature can be as good as a few centimeters. These microbend transducers can be configured to respond to very small mechanical strains, and detection thresholds in the region of a few tens of microstrains over a few meters of gauge length are typical.

There are numerous examples of practical installations of distributed strain sensors based upon this principle, typically with total lengths extending to a few hundred meters and spatial resolutions of a few meters.

This approach is, of course, an indirect one in that the strain is transformed to a loss through an independent strain-to-microbend converter. The advantage in this approach lies in the fact that these strain converters can be designed to be relatively temperature-insensitive, though typically practical installations use a parallel dummy strain sensor that is not coupled to the structure and extracts the strain signal from the differential of the two sensing structures. The strain-sensing element itself is straightforward to produce in large quantities using conventional cable-making equipment; and the use of multimode optical fiber rather than single-mode, which typifies the delay-based systems, makes for straightforward interconnect and relatively inexpensive source components. Furthermore, the measurand transformer can be made sensitive to other parameters so that the same basic concept can be adapted to a variety of measurands. Microbend sensors will make a significant contribution to structure evaluation technology particularly in the civil engineering context.

3.4.3.4 Strain Sensors Using Nonlinear Optical Interactions

The principal nonlinear interactions that occur in optical fibers and that can be used for sensing purposes have already been mentioned. To recap these are:

- Stimulated Raman scatter, which involves a frequency shift typically of the order of 13 THz and may be used for both temperature and strain sensing;
- Stimulated Brillouin scatter involving frequency shifts of the order of 10 GHz that can also be used for temperature and strain sensing;
- Fluorescence phenomena usually in materials external to the fiber used in both temperature sensing and chemical measurements;
- The optical Kerr effect, which is the only one of these phenomena not as yet (1995) exploited commercially in sensing but which has demonstrated potential sensitivities to both strain and temperature.

In common with most transduction mechanisms, these interactions all demonstrate some dependence upon both temperature and strain so that simultaneous strain and temperature measurements and/or compensation techniques are essential. However, it is indicative of the power of these approaches that commercial systems are already well developed.

Simulated Raman Scattering (SRS) is the most established of these commercial systems with several manufacturers offering distributed temperature probes based upon this principle. The idea is very straightforward (Figure 3.32). This shows the up-shifted in frequency (anti-Stokes) and down-shifted (Stokes) lines induced through Raman scatter around the optical carrier frequency. The peak of the Raman gain curve in silica occurs at about 13 THz away from the carrier, but the gain itself is significant over a range extending to 40 THz from the carrier reflecting the fact that the Raman scatter process is extremely fast. The statistics of the scattering process is such that the intensities of a scattered power at an energy ΔE above the carrier frequency is related to its mirror image through

$$\frac{I_{\Delta E}}{I_{-\Delta E}} = e^{-2\Delta E/kT} \tag{3.16}$$

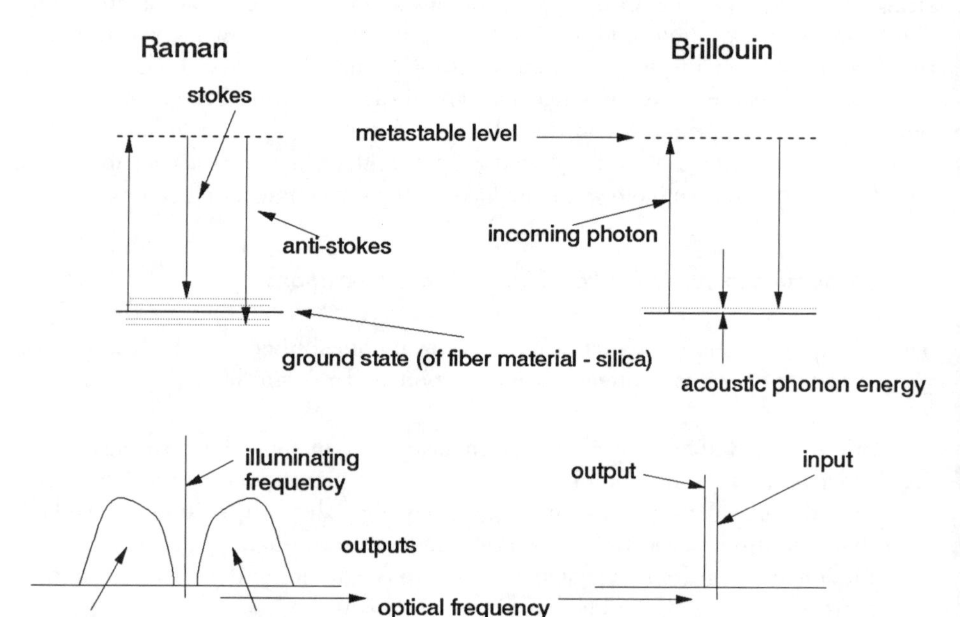

Figure 3.32 Nonlinear scattering processes in optical fiber for use in sensing.

This equation forms the basis of the Raman scatter thermal probe. ΔE is fixed, so the ratio of these two intensities is a function *only* of the local temperature. Even though the location and size of the peak in the Raman gain curve may also vary with both temperature and strain, this intensity ratio remains unaffected by these fluctuations.

Figure 3.33 is a schematic of the distributed temperature-sensing system based upon Raman scatter. The signal-processing system extracts the ratio shown in (3.16) and inverts this to find temperature. There are some slight corrections to this in that the differential attenuation between the Stokes and anti-Stokes wavelengths must be taken into account. To a good approximation, the attenuation process in fibers is dominated by Rayleigh scatter, so a suitable fourth-power correction is all that is required. The value of ΔE corresponding to a 10-THz offset is of the order of 40 meV. At room temperature kT is about 25 meV, so the intensity ratio to be measured is typically in the range 10 to 100.

The principal difficulty involved in obtaining a good value for this ratio lies in measuring its value sufficiently accurately. The power in the Raman backscatter signal is several (typically 5) orders of magnitude below the pulse amplitude used to excite the scatter. Therefore, either long integration times or high pulse power levels are required. Current state-of-the-art performance for these instruments will achieve temperature resolution of the order of 1°C in parallel with the spatial resolution of the order of a meter in integration times of the order of one minute. Future improvements to these instruments, which are presently available from several manufacturers worldwide, will focus upon the source and detection technologies and are likely to involve the use of either very high powered laser diode arrays or optical fiber lasers and amplifiers with the latter offering some potential as receivers as well as transmitters.

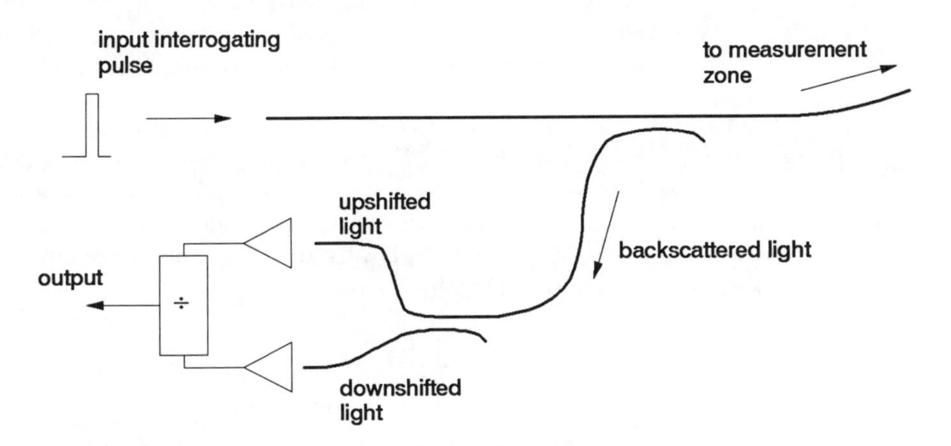

Figure 3.33 Raman-distributed temperature probe: basic schematic.

Raman backscatter has also been used in strain measurement though with much less success. The strain transducer for Raman systems is usually an organic crystal, typically a polydiacetylene addressed using a fiber to illuminate it. The detection technique relies upon measuring the frequency of the peak Raman scatter. The practical applicability of this approach to localized strain sensing using optical addressing remains to be demonstrated. Functionally, this technique is very similar to Bragg grating systems, which for point sensing in multiplexed arrays, appear to offer significant benefits.

Raman scatter in optical fiber is a very forgiving phenomena. It can be excited at all wavelengths and produces a broadband scattering signal that may also be used as the foundation for an amplifying system in the Stokes region. The phenomenon is extremely fast and so is compatible with short-pulse excitation; on a Raman timescale, pulses of nanosecond duration are effectively quasi-static. In essence, this is a reflection of the very broad band response of the scattering process.

Stimulated Brillouin Scatter (SBS), while similar to SRS, is operationally very different. The linewidth for SBS is of the order of 20 MHz, and the frequency shifts are of the order of 10 GHz depending on the wavelengths. The linewidth here is a direct reflection of the phonon lifetime in silica, and it is intuitively reasonable that the linewidth of the laser source used to excite SBS should be less than that of the scattered radiation.

There are other contrasts between SRS and SBS. The necessary power levels for SRS are typically 1 to 2 order of magnitude higher than for SBS. SRS is, for most pulse durations, a function essentially of the peak power in the pulse (that is, the SRS decay time is very rapid); and the coherence of the source, again due to the rapid decay times, has little influence upon threshold powers. SBS in contrast occurs at very low average powers (as low as a few milliwatts) but is very dependent upon the coherence length of the source and the threshold power for SBS increases as the pulse length decreases. Indeed to all intents and purposes SBS disappears for pulse widths less than 10 nanoseconds and is quasi-CW for pulse widths in excess of one microsecond.

The SBS interaction is best understood as coupling between the stress wave set up by the traveling light beam (which since the stress is proportional to the square of the field has a wavelength half that of the optical wavelength) and an acoustic phonon with identical *wavelength*. The optical frequency of the scattered light is shifted in effect due to the Doppler shift imposed by the traveling sound wave through which the backscatter occurs and is given by:

$$f_{\mathrm{Brill}} = \frac{2nV_{\mathrm{a}}}{\lambda_0} \tag{3.17}$$

where n is the effective index of the optical guided wave, λ_0 is the free space optical wavelength, and V_{a} is the acoustic velocity in the fiber. This acoustic velocity has two components due to the bulk wave velocity in the silica from which the fiber is fabricated: V_{bk} and the stress-induced velocity components V_{st} that depend upon the tension applied to the fiber:

$$V_a^2 = V_{bk}^2 + V_{st}^2 \tag{3.18}$$

The acoustic wavelengths concerned are small compared to the dimensions of the fiber, so the acoustic propagation is dominated by the bulk component. The stress component depends upon the square root of the applied tension and under these conditions can only impose a small perturbation to total acoustic velocity. Under these circumstances

$$V_a \approx V_{bk}\left[1 + \frac{cT}{2V_{bk}^2}\right] \tag{3.19}$$

where c is a constant, and T is the applied tension (in proportion to the applied strain). Consequently, the change in frequency shift given by (3.17) is proportional to any applied tension. Note that there are also implicit thermal components (the material parameters that determine V_{bk} include the Young's modulus and the density, and there are often thermally induced changes in stress) and also an inverse dependence of the offset frequency on the wavelength of the laser light. Therefore, any measurement system must have a stabilized optical frequency and/or some from of referencing.

The Brillouin system has been significantly researched, and Figure 3.34 shows a simplified block diagram of a commercial system used to detect strain as a function of position along an optical fiber. Typical values of the strain and temperature

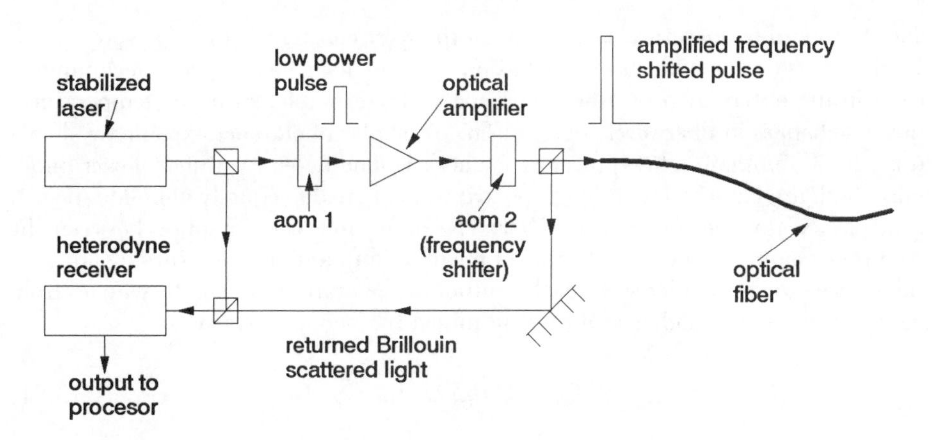

Figure 3.34 Schematic diagram of basic Brillouin heterodyne OTDR.

frequency shift coefficients at a pump wavelength of 1.5 μm are 1 MHz/°C and 5 kHz/μstrain operating around an offset frequency f_{bf} approximately 13 GHz. The resolution that can be obtained depends upon the spectral width of the Brillouin-shifted radiation and the receiver signal:noise ratio. The effects of phonon lifetime influence the longitudinal resolution available for a particular strain value to the order of 10m (corresponding to 100-ns round-trip delay), and practical limitations to the signal:noise ratio imply a strain sensitivity of 100 μstrain over this length. In *distance* terms this corresponds to a change in the length of a 10-m section of 1 mm due to applied strain. The thermal crosstalk is such that 100-μstrain signal corresponds to a temperature change of the order of 1°C.

The Brillouin system, particularly when combined with effective temperature compensation, is a powerful tool for distributed strain monitoring within its operational framework. There appear to be fundamental reasons why spatial resolutions much less than 10m are unlikely to be achieved; similarly, without very large increases in received signal level, the 100-μstrain threshold will be difficult to improve upon particularly when the cross-temperature sensitivities are accommodated. Additionally, short sections subjected to rapidly spatially-varying strain fields will be difficult to resolve since the individual section lengths will be inherently too short to build up a significant SBS signal. On the other hand, the SBS system will function over fiber lengths greatly in excess of 100 km, and it has obvious utility in the context of cable monitoring and testing and for use in very large structures. Commercial systems with this application in mind are becoming available.

The *optical Kerr effect* is the intensity-dependent contribution to the refractive index and is denoted by

$$n = n_0 + kI \tag{3.20}$$

The strain and temperature coefficients of the Kerr constant k are relatively low, so the Kerr effect to date has been used as an optically controllable probe technique to assist in the observation of other phenomena, for example, strain or temperature-induced changes in fiber birefringence. The principles of one such system are shown in Figure 3.35. Again a low-power probe laser is modulated by a high-power pump pulse. Both are launched with linear polarization aligned to equally populate the two principal axes of the birefringent fiber. The pump produces coupling between the two probe channels (one in each state of polarization), and due to a varying effective input phase to the coupler section, the output at the probe wavelength switches from one polarization to another at a predetermined frequency given by

$$f_{\text{Kerr}} = C_f B_f \Delta\lambda / (\lambda_1 \lambda_2) \tag{3.21}$$

where c is the velocity of light in the fiber and B_f is the birefringence of the fiber.

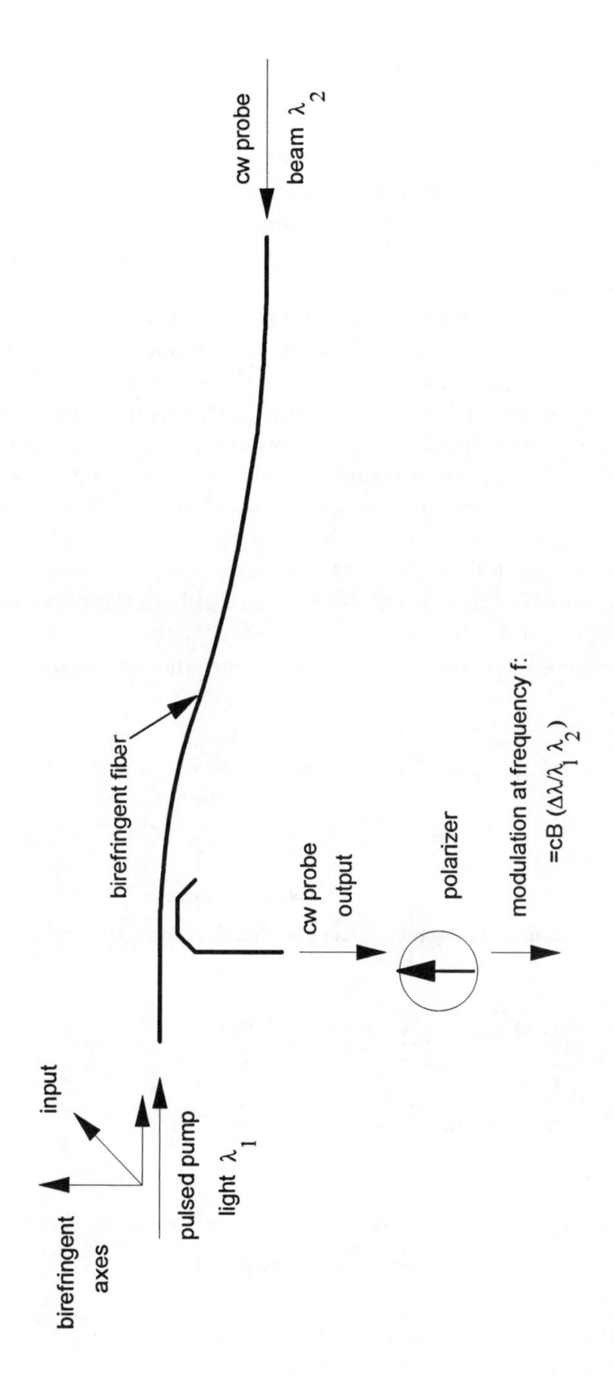

Figure 3.35 Use of the Kerr effect to probe fiber birefringence *B* into modulation frequency *f* by injecting a pump pulse (*c* is the velocity of light in the fiber).

This frequency, f_{Kerr}, is a function of C_f and B_f, assuming that the pump and probe wavelengths are kept sufficiently stable. Both these parameters depend on temperature and strain, so the usual crosstalk problems manifest themselves. This does have the potential to derive a frequency output in a distributed sensing system. However, since the frequency signal evolves due to traveling interactions between the pump and probe waves, the achievable spatial resolution is of the order of the inverse of the wavelength f_{Kerr} when measured as a subcarrier frequency on the optical signal on the fiber.

Distributed strain and temperature sensing using the Kerr effect has yet to be properly demonstrated even at the laboratory level; and factors such as strain sensitivity, dynamic range, signal:noise ratio, spatial resolution, and thermal crosstalk have yet to be resolved. The achievable technical performance appears to be no better and possibly worse than that offered by the Brillouin system, and furthermore, the Kerr system needs special fibers and similarly complex source and detector electronics and optics. To ascertain where these interaction mechanisms offer any significant benefits will require further detailed research.

There are many *fluorescence*-based optical fiber sensors, though to date none have been configured in a distributed format; although the concept of a fluorescent cladding, Figure 3.36, has been mooted. The spatial resolution of such a system would be limited by the fluorescence decay times that are typically microseconds and

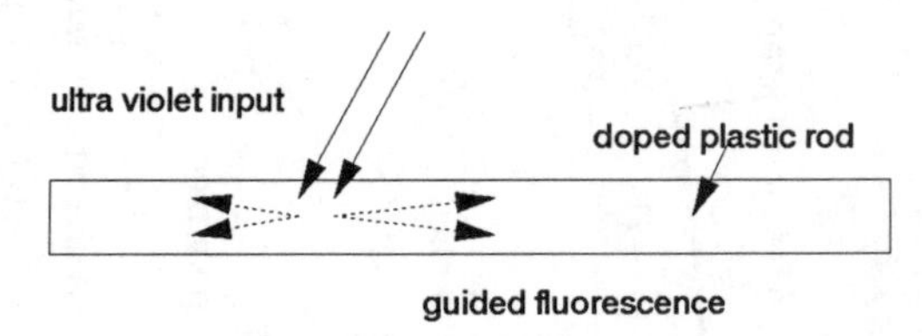

(a) ultra violet to visible conversion in doped plastic guide

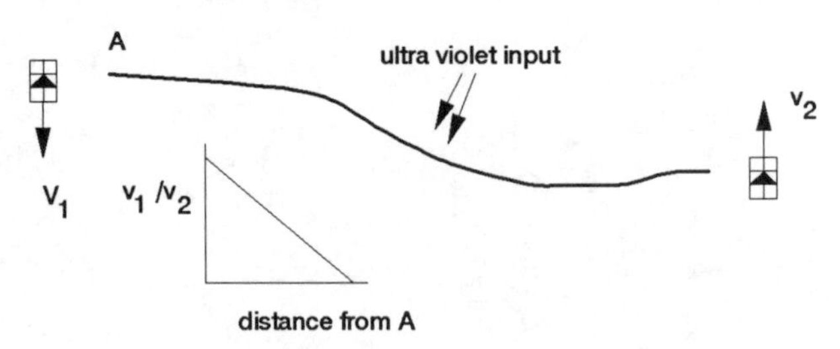

(b) rhodamine 6G-filled fiber version

Figure 3.36 The use of fluorescent fibers as distributed sensing elements: (a) ultraviolet to visible conversion in doped plastic guide; (b) rhodamine 6G-filled fiber version.

could be longer. The more usual configuration for fluorescence-based systems is a point sensor. The most established of these is a temperature probe using a rare earth phosphor and examining the change in the fluorescence spectrum as a function of temperature. This device has been available for several years as a commercial product and has made its mark in temperature measuring in environments in which electrical systems are totally unacceptable, for example, microwave cooking and radio-frequency drying. However, the cost of the instrument remains prohibitive for use in these less technically demanding structural monitoring applications, especially since here arrays of sensors are usually required.

3.4.3.5 Fiber Optic Sensors in Structural Monitoring—Some Observations

Physical parameter sensing using optical fibers is now relatively well established, and in particular the crosstalk between signals produced by temperature and strain fields has been recognized and accommodated. Optical fiber technology is particularly attractive since it permits distributed measurement, thereby enabling parameter variations as a function of position to be evaluated throughout a structure. While there are a number of systems commercially available, there remain several important technical issues associated with the monitoring of relatively small structures with adequately high resolution and the necessary temperature stability. There are also significant issues, some of which will be addressed in Chapter 6, concerning the applications engineering of these systems into structural monitoring; in particular, environmental compatibility and the relative merits of point, distributed, and quasi-distributed sensing must be addressed.

3.4.4 Other Techniques

The science (or art) of structural monitoring continues to attract the ingenious and innovative solution. The challenge is always to find simple, cost-effective ways of addressing large-area structures, preferably with the ability for multipoint or distributed measurement, and—mandatory—without using complex, heavy, and interference-susceptible wiring harnesses.

One of the more interesting of these concepts to emerge in recent years is the radio-addressable sensor patch shown in Figure 3.37. The idea is that a resistive strain gauge (or it could be a temperature gauge) is built into a flexible surface-mounted (or embeddable) patch. The environmentally modulated resistance is built into a RLC circuit that is interrogated through independent inductive coupling. The effective impedance of the inductively coupled circuit is a function of frequency and of the resistance and the relative location of interrogation and receiving coils—which can be fixed. Neural network processing can in turn give the optimum conditions of

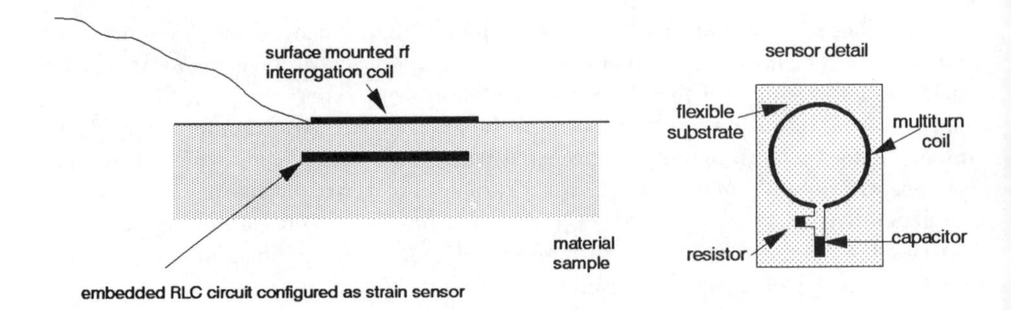

Figure 3.37 Embedded sensor technology using inductively coupled mechanically flexible resonant circuits.

frequency and drive voltage for best measurement of the parameter of interest (strain or temperature).

Initial trials gave results [20] indicating that strain could be monitored to an accuracy of the order of 1% of range for the patch embedded within a glass epoxy composite. The structure has the advantage of low cost, material compatibility, simple construction, and simple monitoring techniques. The need for neural network processing of the interrogating and received signals is not onerous given low-cost computing. However, the technique in its present form is essentially off-line and only permits point measurements. Regardless, its overall simplicity is persuasive and will ensure its utility in applications where these attributes do not compromise the measurement.

Many of the more conservative users of structural monitoring systems make expensive use of the vibrating wire gauge. This simply converts strain in a structure to the resonant frequency of a steel wire mounted in intimate contact with the structure. The wire is mounted between the poles of a permanent magnet and plucked by a current impulse. The induced current is at the vibrational frequency and is readily measured to give strain resolution in the tens of microstrains region.

3.5 CHEMICAL AND BIOCHEMICAL SENSING IN STRUCTURAL ASSESSMENT

3.5.1 Introduction

Chemical and biochemical sensing technology is probably the most fragmented of all measurement technologies. Many of these techniques have some potential in the smart structures domain. To date relatively little experience has been accumulated in the use of chemical and biochemical systems in structural assessment. The diversity of chemical systems may be appreciated from the observation that they cover everything from the measurement of toxins in effluent to corrosion in steel and polymers to dust and gas monitoring in the atmosphere to blood gas assessment.

In the structural assessment regime there are two principal applications areas:

1. Safety, environmental, and legislative requirements that impose the need for detection of toxic, inflammable, or otherwise hazardous contaminants;
2. Corrosion assessment and the detection of the conditions under which corrosion occurs.

The discussion on chemical sensing that follows can be constrained a little by limiting its scope to systems that are capable of on-line measurement within the structural environment and that can monitor chemical content over reasonably wide areas. Even so, we need to select from hundreds of techniques. Here we shall very briefly consider the basic features of electrochemical sensors, chemical-to-mechanical conversion phenomena, and spectroscopic techniques, the aim being to present an impression of the range of options that is available.

3.5.2 Absorptive Chemical Sensors

By far the majority of gas-sensing devices rely upon selective (or nearly so) absorption of the gas of interest into some form of electrode structure. This absorption typically ionizes the gas concerned and produces a resultant change in conductivity within the electrode structure. There is a bewildering variety of such sensors: Ohba lists about 100.

The most common of these systems rely upon either changes in the resistivity of metal oxides or changes in the electronic potential of the gate electrodes in MOS semiconductor devices.

Toxic gases (for example, carbon monoxide) can be typically detected using doped tin oxide films. These are a well-established, inexpensive, and barely understood technology that is widely used in point sensors—including battery-powered ones—for domestic and industrial applications.

Inflammable gases are usually detected by exploiting selective catalysis based upon palladium or platinum electrodes. These typically ionize the concerned gas, producing ignition, enhancing its chemical activity, and changing local electrical properties. In some versions—most notably the ubiquitous pellistor—the catalyst is heated and induces local ignition behind the flame proof barrier, thereby introducing drastic changes in local resistivity. Pellistors are sold by the million for use in safety monitoring systems.

Semiconductor devices exploiting these phenomena usually invoke the modification of the gate structure in an MOS transistor. These devices can obviously be fabricated at the wafer level, but their advantages only come to the fore when multiple-element systems using several different gate electrode structures on a single wafer are required. For single-channel structures there are no real cost advantages in the semiconductor technology where signal and power cables and safety legislation often dominate.

It is obviously axiomatic that all these devices involve electrode structures that are exposed to the atmosphere and, by their very nature, are chemically active. Their low cost and simplicity is then somewhat tempered by a tendency toward poisoning and a deterioration in sensitivity with age. Consequently, many of these devices impose a relatively high cost of ownership since they need regular replacement. Additionally, the heated catalytic devices (most notably the pellistor) require high current drive with subsequent cabling costs that are especially problematic in intrinsically safe areas and also impose practical limitations on the distance over which the pellistor may be interrogated, typically to the order of one hundred meters.

There is a bewildering range of electrochemical transduction techniques. Within this range perhaps the most important are chemically activated semiconductor and electroresistive devices. These all operate by selectively absorbing typically several molecular species. This implicit cross sensitivity problem has been approached by using an array of different devices together with intelligent processing using a neural network approach to provide a best guess at the species present. These basic principles underlie a whole generation of "artificial noses" that are beginning to emerge as preproduction prototypes from a number of U.S. and European companies. The secret in such devices lies within the chemistry of the interaction zone and in applying appropriate signal correction to the signals from each element to allow unambiguous identification of a particular species. The detection system must also correct for the fact that the sensitivity and relative sensitivities of a particular sensing element to the various species being addressed varies, often significantly, with temperature. These devices have yet to make their impact as a contributor to structural instrumentation technology but certainly may be regarded as "smart" especially when interfaced to electronic processors.

One of the most important electrical sensing techniques for structural instrumentation is also one of the simplest: the use of electrical time domain reflectometry to address a cable array. This typically comprises a multiplicity of point sensors within an array. The sensor cable impedance varies depending upon whether the medium surrounding the cable is wet or dry. These TDR probes are in extensive use as point location moisture ingress measurement systems, as moisture level monitors in soils for agriculture and horticulture, and in protection systems as fluid alarms for, for example, computer installations and precious archives [21].

In principle (Figure 3.38), electrical TDR is similar to its optical cousin: Indeed, it has been available for a much longer period of time. As a measurement tool, its limits are imposed by the losses and dispersion characteristics of the copper-based cable. For most TDR systems the total range rarely exceeds 1 km. Further attenuation in copper cables increases rapidly with frequency, so precision length measurements are precluded.

Most TDR-based sensing systems operate over very short (tens of meters) ranges and are capable only of single point ingress (for example, of moisture) resolution. Often the probes are only a meter or so in length and used for moisture detection.

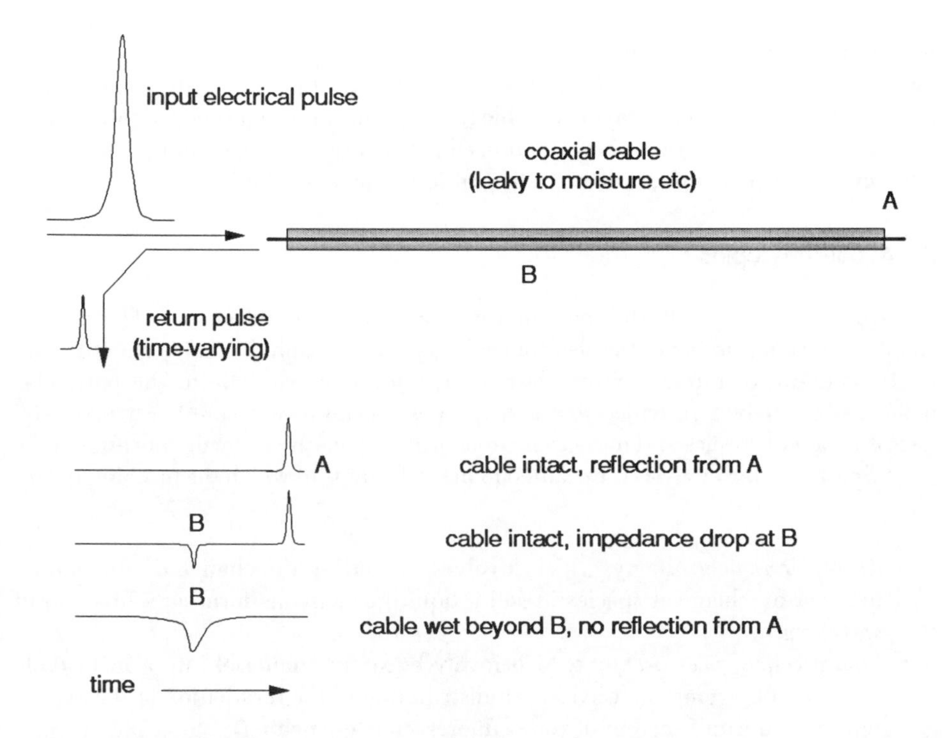

Figure 3.38 Electrical TDR usually uses coaxial cables, or—even looser—central conductor probes. Ranges for cable measurement are a few kilometers and for dielectric measuring probes as low as a few meters, with only one useful resolution point.

The system can, however, be very simple, so that a most cost-effective measurement and monitoring system, especially for moisture, can be realized.

3.5.3 Chemimechanical Sensors

Many materials undergo a mechanical change in the presence of one or more interactive chemical species. Of these changes, physical size, hardness, stiffness, and density are probably the most useful. Examples of these processes include absorption of hydrogen in palladium (a highly selective process), water and ion selective absorption into hydrogels, or physical changes induced by corrosion processes. In all these devices the material must be integrated into a mechanical transducer that monitors its changes in mechanical properties, frequently using optical techniques. These systems are also among the very few that may be used to monitor the spatial distribution of a chemical throughout an area of interest; so in principle the chemimechanical transduction process is extremely versatile, though like other chemical sensing systems it almost always exhibits some sensitivity to chemical species other than the one of interest. Also in common with electrical transducers, the chemical-to-mechanical conversion process is not always reversible though some of the most useful (absorption in hydrogels and catalytic action) can indeed be exercised through multiple measurement cycles, provided that adequate desorption time is available.

3.5.4 Spectroscopies

Spectroscopy is probably the most powerful chemical analytical tool. The process enables external probing of the electronic energy levels within the molecular structure of the material of interest. Since these energy levels are unique to the particular molecules of interest, provided that their properties can be measured very precisely, identification of species and measurement concentrations are immediately attainable.

Spectroscopic analysis is exploited in many formats of which the most important are:

- *Absorption spectroscopy*, which involves measuring the change of attenuation induced by chemical species in solid, liquid, or gaseous form as a function of wavelength;
- *Fluorescence spectroscopy*, which involves exciting material into a metastable state and observing the wavelength distribution of the reradiation of the excitation—often as a function of time (fluorescence quenching);
- *Raman spectroscopy*, which entails measuring the spectrum of reradiated emission from a material and comparing this to the incident spectrum. This corresponds to excitation to a metastable state and relaxation to a state adjacent to the ground state. The energy difference between this state and the ground state produces a shift in the spectrum and is also a function of the material under examination.

Spectroscopy in most cases implies optical spectroscopy covering the range from the mid-infrared to the near-ultraviolet. However, there are also useful microwave spectroscopies that typically examine the rotational frequency of molecules and nuclear spectroscopies, which involves subatomic particles as the probes (for example, Mossbauer spectroscopy). These techniques more typically probe differences in nuclear energy levels rather than differences in electronic energy levels.

For optical spectroscopies there are a number of general points that can be made concerning the technique and its interpretation. Of these the most important are as follows:

- In liquids and solids the close proximity of adjacent atoms always implies that any absorption lines are relatively broad with a bandwidth that is typically several percent.

- In gases the molecules are several orders of magnitude further apart and the absorption lines are correspondingly much narrower with bandwidths typically extending over a range of the order of gigahertz.

These observations have a direct effect upon the design of a spectral instrumentation (Figure 3.39). In monitoring systems a remote sample cell is typically required so that optical fiber links will often be incorporated into the design of the system. The spectrometer measures the product of the output of the optical source with the transmission of the cell in the associated system component over the wavelength range of interest. The resolution in wavelength terms of this measurement is typically very different for gas and liquid phase measurements, necessitating different approaches to the spectrometer design process.

In common with other chemical measurement techniques, spectroscopies are utilized in a bewildering variety of formats and instruments, some of which are shown schematically in Figure 3.40. Of these probably the most relevant for structural instrumentation are surface-monitoring probes, which can, for example, yield information concerning the progressive change in surface parameters and are particularly effective in detecting the onset of chemically induced surface corrosion. The surface-measuring probe will also be configured to monitor changes in scatter parameters that in turn can yield information concerning surface morphology. Surprisingly, of the spectroscopic instruments illustrated generically in Figure 3.40, this is the one that has had probably the least attention and has certainly yet to graduate from the

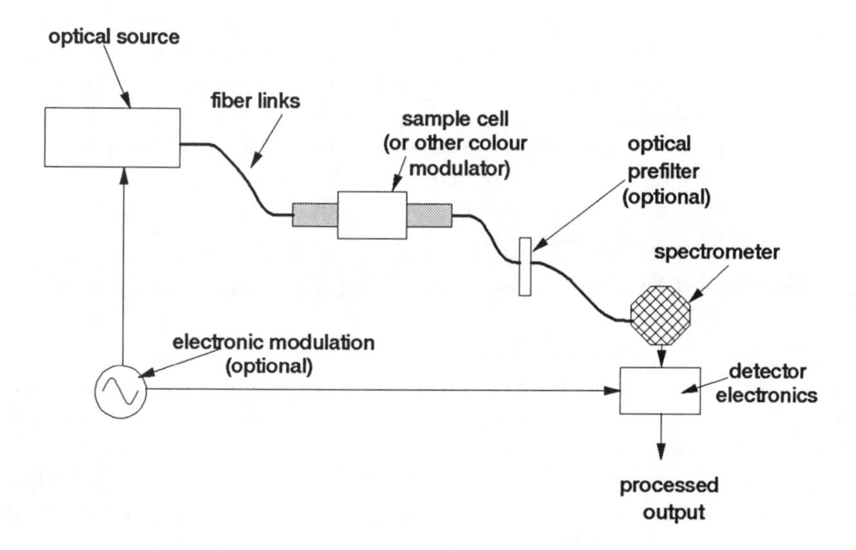

Figure 3.39 Elements of the fiber optic-based color measurement system.

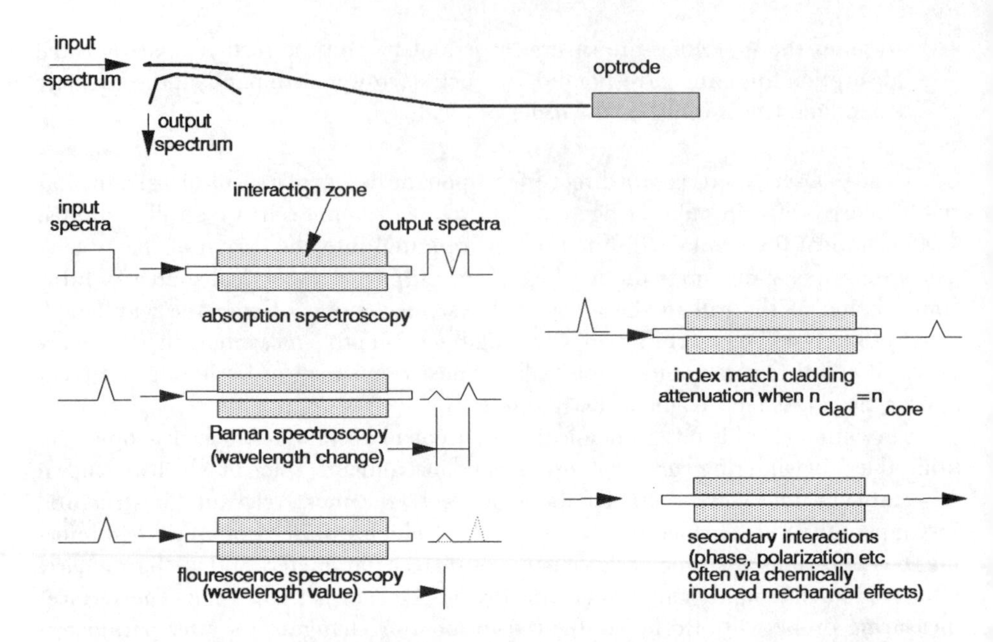

Figure 3.40 Basic principles of most chemically sensitive optrodes. Usually they are used in reflections as shown above. Direct surface reflection may also be used (without the Optrode).

laboratory into field instrumentation. For structural monitoring perhaps the most obvious constraint on such a probe is its inability to effectively cover large areas.

Spectroscopy is a very powerful analytical tool. It operates using materials-specific signatures and examines the sample without the need for intermediate chemistry. It has already had a profound impact upon chemical measurement systems and will continue to do so. However, its use in structural instrumentation has thus far been restricted to airborne measurements of pollutants from chemical plant and similar open path applications.

3.5.5 Fiber Optic Chemical Sensing Systems and Distributed Measurements

The vast majority of chemical measurement systems are either optical or electrochemical and, with the exception of direct spectroscopy, all involve intermediate chemical stages with the consequent possibilities for interference from unwanted measurands and/or from thermal effects.

Optical measurements are almost entirely spectroscopic—though with some important exceptions. The move from bulk optics to guided wave optics results in the familiar benefit of rugged compact probe with the capacity for interrogation over a considerable distance through an electromagnetically robust link.

Sensing systems linking optical fibers to color-sensitive chemical measurement devices are gaining wide acceptance. The design of the sensing head—often dubbed the *optrode*—continues to attract research effort exemplified through annual international conferences devoted to the subject.

The exploitation of optrode systems in structural monitoring is limited primarily because these devices are point sensors and are difficult to multiplex effectively. They can, however, efficiently detect the presence or otherwise of corrosion products or corrosive conditions, and their small size and flexibility in implementation may be beneficial. Multiplexing optrodes has not, to this author's knowledge, been seriously addressed other than through a switched source system. The reasons are simple to identify. Signals from optrodes are often relatively weak since many rely upon collecting a sample of scattered light; and the chemical processes involved can be slow, producing long decay times in returned signals and therefore inhibiting the prospects for the most common time division-based multiplexing systems. Perhaps also the emphasis on optrode research has been in the medical and environmental monitoring areas where the time constants of the phenomena of interest are relatively slow and also where samples from an individual point present very useful data. A drive for large multiplexed arrays in a structural monitoring context could stimulate an examination of alternative optrode-compatible multiplexing systems—of which there are very many.

Much has already been made of the benefits offered by optical fibers as distributed measurement systems. However, to date there has been relatively little progress made in the application of these distributed concepts to chemical measurement. There are three generic approaches (Figure 3.41) that have been investigated to date:

- The use of a chemically sensitive coating or indeed a chemically sensitive fiber as the transduction element;
- Exploitation of evanescently coupled direct spectroscopy;
- Incorporation of a chemical:mechanical transduction system and detection of the resultant chemical changes using standard optical fiber techniques.

Coatings are historically the best established chemical-to-optical transduction systems and indeed form the basis of many optrode designs. Typically the active material is incorporated with a polymer or sol-gel host of lower index than the fiber core and the signal is generated through evanescent field penetration of the light propagating in the typically multimode fiber into the active cladding. The homogeneity of this evanescent field penetration depends upon the mode spectrum in the interaction zone, and collection of the modulated light is usually inhibited by the fact that this light is often homogeneously scattered by the modulation process. The attenuation in the modulation zone is typically quite high, often measured in decibels per centimeter, so the maximum interaction length is typically less than 1m. This,

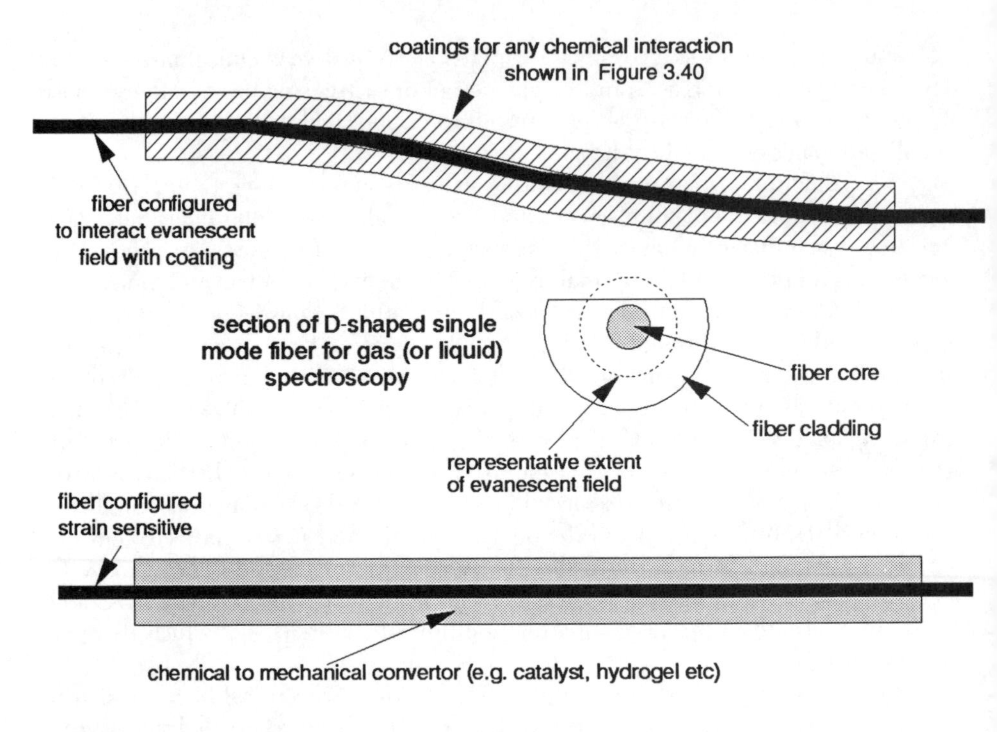

Figure 3.41 Principal techniques for distributed chemical measurement.

of course, constrains the potential of such techniques for distributed measurements, though a few systems based upon short interaction lengths spaced at regular intervals along the fiber have been reported.

Direct evanescent wave spectroscopy using a D-shaped fiber (Figure 3.41) can be configured for use in both liquid and gaseous host media. The principal features of this system are that the optical fiber transmission medium is single mode and that careful control of the core:flat distance determines the penetration of the evanescent field. An accurate knowledge of the refractive index within which the fiber is to operate is also essential to control the penetration of the evanescent field into the sample and therefore the measurement sensitivity. Penetration depths in aqueous media are significantly higher than those in gaseous media. Evanescent devices may be configured to operate in direct spectroscopic mode or to excite local fluorescence, though for the latter the collection efficiency of the fluorescent radiation is often very low. Since fluorescence is intrinsically a slow process with a long decay time, fluorescence-based systems are usually unsuited to incorporation into distributed sensing: The distance resolution is directly related to the fluorescence lifetime, which can extend into the microsecond region.

For direct absorption spectroscopy the situation is very favorable, the process is electronic and intrinsically very fast, and the evanescently coupled structure can be arranged to provide attenuation in the presence of, for example, methane gas, of up to 1 dB/m at 100% concentration—or equivalently 0.01 dB/m at 1% concentration (that is, 20% of the lower explosive limit of methane gas).

State-of-the-art OTDR interrogation units are capable of resolving attenuation changes of approximately 0.01 dB/m with submeter resolution. For gas-detection systems it is fundamental that the gas disperses so that resolution of the order of meters is required in any distributed measurement. Consequently, it is apparent that this basic system could, in principle, resolve a 0.01-dB change in attenuation over a length of a few meters. This is adequate to respond to the safety limits for the detection of methane [22].

In the structural-monitoring context, methane is a frequent hazard, especially in gas and oil, mining, distribution and processing, and in water plant and in agricultural buildings. It is, however, rarely indicative of structural corrosion. For this last application a liquid-based sensing system is more appropriate, exploiting evanescent wave coupling into an aqueous medium with all the threats to the mechanical integrity of the fiber optic which that implies. Perhaps some intermediate chemistry—maybe an indicator immobilized to the sol-gel film—would be appropriate, but these techniques have yet to be explored.

The third category of distributed measurement (Figure 3.41) involves the use of chemical-to-mechanical transformers. This technique—while still not yet commercial—has thus far shown the most promise in the distributed detection of chemical parameters for structural assessment. The structure shown in Figure 3.41 uses a hydrogel compound as the interface between the chemical of interest and a microbend transducer that in turn modulates a loss through the optical fiber. This interface, in its simplest form, responds to moisture. Hydrogels may swell by up to several hundred percent when completely wetted. The hydrogel can, however, be tailored to switch in response to other parameters such as pH and some ionic species, so this is a natural candidate for the assessment of corrosion conditions in structures. An initial trial (Figure 3.42) demonstrated good spatial resolution for the presence or absence of water and successful identification of an aqueous grout in a reinforcing tendon duct during blind tests (see Chapter 6).

In its present form the hydrogel-based sensing systems are switches that trigger on the presence or absence of a particular chemical species. Future activity will expand the range of species to which the hydrogels investigate the prospects for low-resolution analogue measurement, for example, of moisture content or of pH over a limited range.

There are other mechanical:chemical transduction process of which catalytic participation is probably the most common. This has been used in point sensor for hydrogen but has not yet been expanded into distributed chemical measurement—though perhaps it should be.

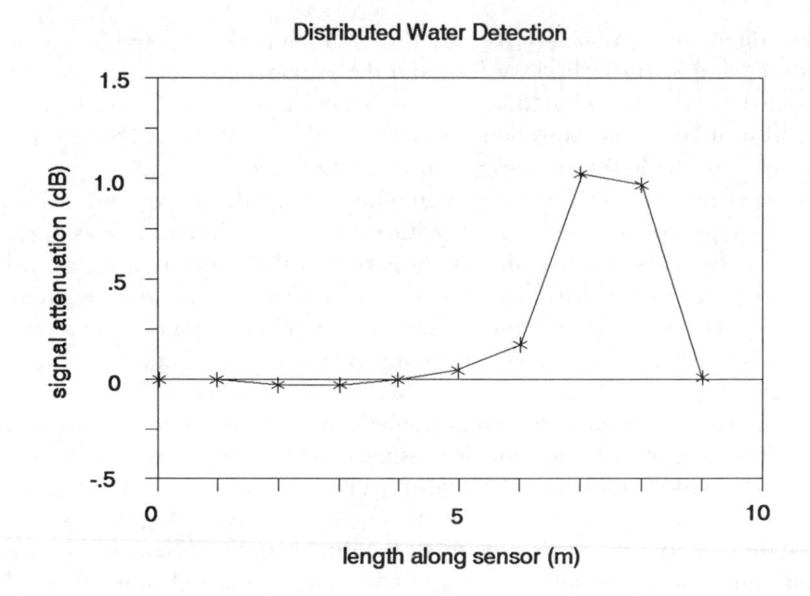

Figure 3.42 Distributed water ingress monitoring system—specimen results.

3.6 SENSOR SYSTEMS—CONCLUDING COMMENTS

It is difficult—even for an optimistic altruist—to attempt to encapsulate the sensor technologies that may be applied to structural assessment in a single chapter. However, in the preceding pages I hope to have encapsulated the principal criteria that should be applied to the selection of sensor systems for structural assessment and reviewed some of the principal technologies. The criteria are relatively simple.

The principal parameter in structural assessment is invariably corrosion, though the application of unusual loads that may stress the structure beyond its limits should also be monitored. Consequently, the following comments apply:

- Strain and temperature measurements can give an indication of loading conditions or, if the loading conditions are known, can indicate the effect of fatigue and corrosion on the structure.
- Chemical assessments may indicate the onset of corrosion conditions and corrosion products.
- In all cases the structure should be monitored using a strategically located array of sensors of a distributed sensor.

This last point is particularly important, especially when new mechanical material systems are considered, since many sensors are mechanically less robust than the materials that they are designed to monitor. It is here that optical fiber-based systems (admittedly a prejudice of the author) come to the fore, and this coupled with the unique ability of optical fibers to perform distributed measurements has skewed the interest in structural sensing toward the fiber optic technology. That said, there are many other alternative sensors that are often more appropriate. For example, piezoelectric films provide good strain-sensing potential over fairly wide areas and electrochemical cells are very useful in monitoring corrosion products.

Finally the role of signal processing in handling data from sensor arrays should not be understated. It is almost invariably essential to apply "intelligence" to the recovered data. For example, it is frequently only trends in this data rather than detailed measurements at a particular time that give the desired information concerning structural integrity. Further, for many sensor systems (especially but not exclusively chemical systems) array data must be interpreted from not quite orthogonal sets to resolve cross-sensitivity issues. The benefits that can accrue from low-cost and very powerful computing systems in this context are an essential feature in applying sensor technologies to structural assessment.

REFERENCES

[1] Ohba, R., *Intelligent Sensor Technology*, Chichester, UK: Wiley, 1992.

[2] Middlehoek, S., and A. C. Hoogerwerf, "Classifying Solid State Sensors: The Sensor Effect Cube," *Sensors and Actuators—State of the Art of Sensor Research and Development*, Lausanne: Elsevier Sequois, 1987.

[3] See, for example, Kittel, C., *Introduction to Solid State Physics*, Fifth Edition, New York: Wiley, 1976.

[4] Moulson, A. J., and J. M. Herbert, *Electroceramics—Materials, Properties Applications*, London: Chapman and Hall, 1992.

[5] Jaffe, B., W. R. Cook, and H. Jaffe, *Piezoelectric Ceramics*, London: Academic Press, 1971.

[6] Lovinger, A. J., "Ferroelectric Polymers," *Science*, Vol. 220, 1983, pp. 1115–1121.

[7] Blom, F. R., D. J. Yntema, F. C. M. van de Pol, M. Elwenspoek, J. H. Fluitman, and Th. J. A. Popma, "Thin Film ZnO as a Micormechanical Actuator at Low Frequencies," *Sensors and Actuators*, Vol. a21-a23, 1990, pp. 226–228.

[8] Chiarelli, P., D. de Rossi, and K. Umezewa, "Smart Polymeric System for Electromechanical Transduction," *Proc AGARD*, cp 531, paper 14, 1993.

[9] LVDTs are standard items from several transducer manufacturers: Application notes from, for example, RDP Electronics Ltd will provide extensive information.

[10] Culshaw, B., and P. T. Gardiner, *Smart Structures—The Role of Fibre Optics*, NATO Lectures on Smart Structures and Materials, 1993, and *Fiber and Integrated Optics*, Vol. 2, 1993, pp. 353–373.

[11] Udd, E. (ed.), *Fibre Optic Smart Structures*, New York: Wiley, 1993.

[12] Culshaw, B., and J. P. Dakin, *Optical Fibre Sensors Vols. I & II*, Norwood, MA: Artech House, 1988 & 1989.

[13] Udd, E., *Fibre Optic Sensors: An Introduction for Engineers and Scientists*, New York: Wiley, 1991.

[14] Dakin, J. P., D. J. Pratt, G. W. Bibby, and J. N. Ross, "Distributed Optical Fibre Raman Temperature Sensors Using a Semiconductor Light Source and Detector," *Electronics Letters*, Vol. 21, 1985, pp. 569–570.

[15] Shimizuu, K., T. Horiguchi, and Y. Koyamada, "Measurement of Distributed Strain and Temperature

in a Branched Fibre Optic Network by Using Brillouin OTDR," *Proc OFS(10)*, Glasgow, Scotland, Oct. 1994 (*SPIE*, Bellingham Washington, Vol. 2360), p. 142.

[16] Handerek, V. A., F. Parvaneh, and A. J. Rogers, "Analysis of Frequency Derived Measurement of Linear Birefringence in Optical Fibres," ibid, p. 588.

[17] Gusmeroli, V., C. Mariotti, and M. Martinelli, "Absolute and Simultaneous Strain and Temperature Measurements by a Coherent Optical Fibre Sensor," ibid, p. 199.

[18] Bhatia, V., M. E. Jones, J. L. Gracem, K. A. Murphy, R. O. Claus, J. A. Greene, and T. A. Tran, "Applications of "Absolute" Fibre Optic Sensors to Smart Materials and Structures," ibid, p. 171.

[19] Davis, M. A., and A. Kersey, "Fibre Fourier Transform Spectrometer for Decoding Bragg Grating Elements," ibid, p. 167.

[20] Spillman, W. B., S. Durke, and W. W. Kuhns, "Remotely Iinterrogated Sensor Electronics (RISE) for Smart Structure Applications," *Proc ECSSM2*, Glasgow, Scotland, Oct. 1994 (*SPIE*, Bellingham Washington, Vol. 2361).

[21] See, for example, *Proc. of Symposium and Workshop on Time Domain Reflectometry in Environmental, Infrastructure and Mining Applications*, North Western University, Evaston, IL, Sept. 1994; *SP 19–94*, U.S. Bureau of Mines.

[22] Culshaw, B., "Distributed Gas Sensing, Fact or Fancy?" *Proc. Europto Conference Optics for The Environment and Public Safety*, *SPIE*, Vol. 2507, 1995.

SELECTED BIBLIOGRAPHY

Sensing and instrumentation are continually developing. Conference proceedings issued in the following series are useful.

International Conferences on Intelligent Materials (ICIM);
European Conferences on Smart Structures and Materials;
SPIE Series on Smart Structures and Materials;
Recent Advances in Adaptive and Sensory Materials and Their Applications;
International Conference on Adaptive Structures;
Eurosensor Conferences;
Transducer Series;
International Conferences on Optical Fiber Sensors (OFS);
SPIE Conference on Chemical and Environmental Sensors;
Europtrode Conferences.

Chapter 4

Actuator Techniques

4.1 INTRODUCTION

Actuation is the controlled release of energy.

Usually, the energy released is mechanical and the controller is electrical. The reservoir from which the energy is released is typically electrical, hydraulic, or pneumatic or stored as a hydrocarbon fuel.

There are, of course, many other actuation processes, including electro-optic phenomena, for example a liquid crystal shutter, remotely controlled electrical switching (your infrared security lamp), and the changing color of the chameleon. We could then generalize the definition of actuation to cover a frightening range of options. We shall resist the temptation.

Actuation is well known—it is an integral and essential feature of all mobile engineering structures and is also very common in buildings and bridges. The concepts range from the automatic ventilator in the greenhouse to the sophisticated flaps and rudders on a large aircraft.

Perhaps though, our smart structure aspires to a somewhat more elegant and general solution. The biological model gives some hints. In all biological systems actuation only occurs exactly where it is needed and is controlled through a combination of local and centralized processing systems. In our current engineering systems, actuation is provided at points that are a compromise between the desirable (which is a distributor actuation network) and the achievable (which is an electromagnetic motor or a hydraulic actuator capable of providing significant forces at a point). The biological model also has another very important feature; namely, it transfers actuation energy to the actuation point very efficiently, usually in a chemical form to be burnt by muscular action at the point at which it is needed. This energy release is controlled locally and monitored through local and central sensory systems.

In summary then our current engineering achievements in actuation are limited to providing mechanical motion at a point and usually to transferring the energy required for this mechanical motion to the actuation point from a mechanical (often hydraulic) or sometimes electrical source. In many ways the conceptual evolution of actuation systems is even further behind sensor networks since very little real attempts have been made to configure actuation in the distributed form and far less has been attempted with the aim of optimizing energy distribution and release. Consequently, the response of engineered structures remains rather naive, and arguably our approach to engineering design has historically been constrained by the implicit requirement to respond at a point rather than throughout the system. Perhaps though, we are beginning to work toward the concept of an artificial muscle.

Here we shall examine the basic features and the options open to implement mechanical changes in structures. There are only a couple of basic functions with which mechanical actuation is involved:

- Changes in position (or its equivalent, to shape) for a mechanical load, which usually involves physical work, for example through changing the shape of an airfoil, through applying a brake, or through repositioning a work piece on a machine tool. Often, as we shall see, much of the work performed serves to change mechanical potential energy.
- Changing the stiffness or mass of the structure usually to modify mechanical resonant frequency often via the function of vibration damping, so by implication the actuator must also be capable of absorbing (or causing to be absorbed) energy within the structure of interest.

This discussion has identified the principal features of actuators that are summarized in Figure 4.1. Our system comprises an energy reservoir that, from the perspective of this discussion, is assumed to have infinite capacity and infinite ability to discharge this capacity; in other words, the response of the actuator system is not

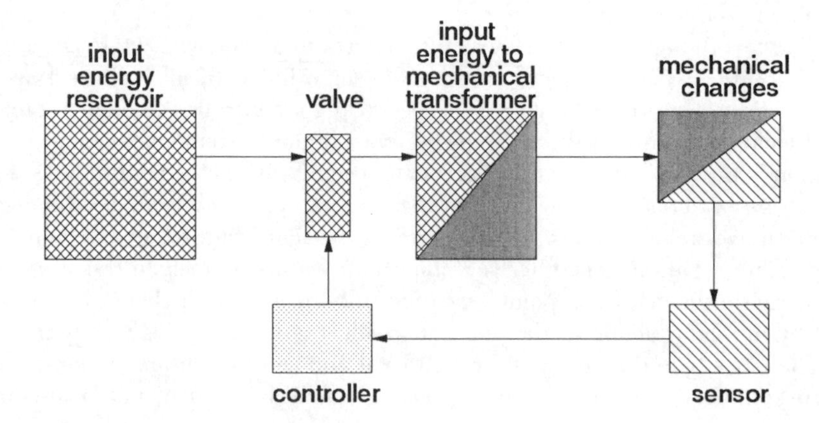

Figure 4.1 The conceptual stages in the actuation process.

limited by the energy source. The release of energy from this reservoir is controlled by a valve that is normally electrically operated. This released energy is then converted into mechanical energy through some form of conversion system. The energy reservoir rarely, if ever, stores mechanical energy. The resulting mechanical change, be it in stiffness or in position, is invariably monitored using some form of sensor system, which in turn is fed back to the control valve. The sensor system is likely to be based upon one of the concepts described in the previous chapter.

The emphasis on actuator functions has broadened over the past decade, probably stimulated through the expansion of the options available for the stored and transmitted energy-to-mechanical work conversion stage, in the associated valve and in the associated sensing system that determines what the actuator has actually done. Also, and in common with sensor systems, the contribution of readily available low-cost, extremely capable signal-processing systems to handle increasingly complex data from sensor arrays and to introduce instructions to increasingly complex actuator arrays has proved to be extremely important. The actuator concept is effectively evolving from a single controlled channel operating a single device (a point actuator) into a multichannel system. This begins to approach the distributed actuator.

The concept has evolved, but, to date, the actuator applications engineering has largely lagged behind the concept. Engineers have yet to implement the distributed actuator concept in a serious fashion, but laboratory demonstrations and engineered prototypes abound. The essential motivation is that the distributed system can put the mechanical energy source in an array of positions where, in principle, the mechanical energy source can cause the greatest effect for the least change in local and total energy. We are essentially here referring to impedance matching. The distributed actuator enables the mechanical drive impedances (about which more will be discussed in the next section) of the mechanical system that provides the energy to be matched into those of the structure that absorbs energy with optimum effect. The potential gains are very significant indeed.

The discussion of actuator systems and the criteria for comparing the potential offered by the various options can be reduced to two principal technical points:

- The efficiency with which energy from the reservoir can be converted into useful mechanical energy;
- The mechanical impedance matching versatility offered by the actuator technology.

There are, of course, a multitude of other parameters, not least of which are installation costs and cost of ownership. These are complex engineering issues. The technical issues are also fairly complex but can be viewed very simply. The conversion efficiency is a function of the actuator technology; impedance matching is a function of the actuator system design. In most cases it is far easier to discuss the former than the latter.

4.2 MECHANICAL IMPEDANCES, CONVERSION EFFICIENCIES, AND MATCHING

It is interesting and useful to digress into a simple account of the mechanical impedance concepts that underlie much of actuator system design and analysis including *both* the *actuator materials* and the *actuator system.*

The simple mass, spring, and dash-pot model of a mechanical system (see Figure 4.2) is surprisingly useful. With careful interpretation of the mass and spring parameters the basic idea can be applied to derive the essential features of the dynamics of almost any mechanical shape or structure. It is especially useful since most intact mechanical structures invariably operate in a linear mode where applied stress produces a proportional displacement.

Using the parameters as defined in Figure 4.2 we have a simple equation of motion for the system:

$$F = m\ddot{x} + \mu\dot{x} + \frac{x}{c} \tag{4.1}$$

This relates to applied force F (usually time-dependent) to the resultant displacement x of the mass from its equilibrium position.

The impedance concept becomes immediately apparent when we recall the equivalent equation for a series resonant electrical circuit:

$$V = L\ddot{q} + R\dot{q} + \frac{q}{C} \tag{4.2}$$

where V, L, R, and C have the usual meanings. The electrical impedance is the ratio of V to $\dot{q}$, so by an analogy we have a mechanical impedance that is a ratio of F to $\dot{x}$. These can be written as Z_{elec} and Z_{mech}, respectively.

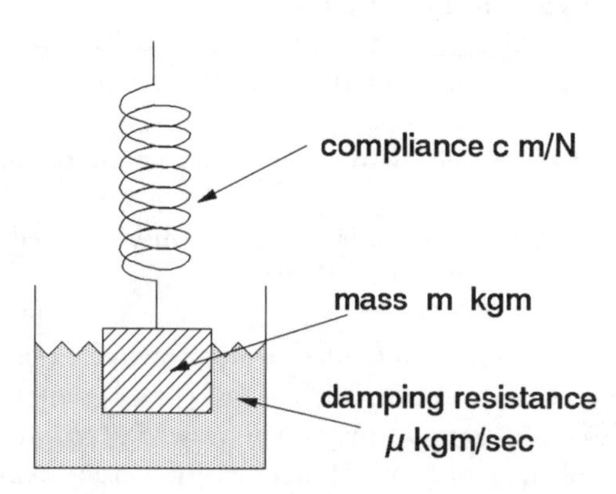

Figure 4.2 Mass:spring:damper model of a mechanical system.

$$Z_{\text{elec}} = j\omega L + R + \frac{1}{j\omega C}$$

$$Z_{\text{mech}} = j\omega m + \mu + \frac{1}{j\omega c} \tag{4.3}$$

The mechanical impedance concept is extremely versatile. Many of the ideas used to excellent effect in electrical circuit analysis and interpretation and also in the definition of electrical materials can be transferred to a mechanical domain and fulfill the same role, such as the following:

- Mechanical energy is stored in the compliance (spring) and mass terms of (4.3). This mechanical stored energy is available for release into a mechanical circuit to produce useful work that is dissipated in the mechanical resistive (friction) term.
- It is useful to distinguish between the mechanical impedance of a structure (equivalent to a mechanical circuit) and of a material (equivalent to a characteristic impedance). For the former, one characteristic often dominates over the others; for example, a spring may be considered as a spring constant (that is, a compliance) neglecting its mass until we approach the self-resonant frequency of the spring. Electrical circuit designers do exactly the same with capacitors.
- The characteristic impedance of the material lies within its *product* of density and stiffness—again analogous to the permeability:permittivity *ratio* in electrical materials—assuming low internal losses.
- The density:stiffness ratio gives mechanical (acoustic) velocities—analogous to the permeability:permittivity product.
- To transfer energy from one part of a mechanical circuit to another entails matching mechanical impedances.

These equivalencies between mechanical and electrical parameters and the very simple basic equations that describe them give us considerable insight into the electromechanical conversion processes that are featured in many actuator functions. At a very general level actuation may be regarded as absorbing controlled (usually electrical) energy and re-releasing this energy mechanically.

In the case where electrical energy is absorbed we arrive at the concept of the *electromechanical coupling coefficient*, which was mentioned in connection with piezoelectric elements in the Chapter 3. This case is of considerable interest since the majority of actuators involves some elements of electrical control. The behavior of any electromechanical system can often be expressed very conveniently in terms of an equivalent circuit. Figure 4.3 shows an equivalent circuit for a piezoelectric element. The key lies in identifying that the series resonant circuit is *entirely* associated with the mechanics of the process with a geometrical capacitance in parallel. The

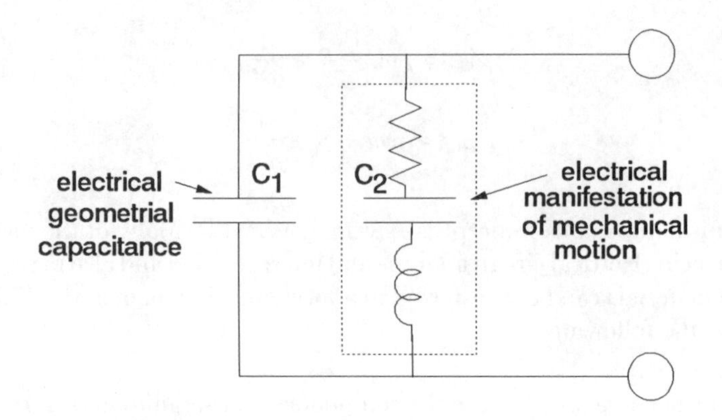

Figure 4.3 Electrical equivalent circuit for piezoelectric transducer. Similar circuits can be derived for, for example, electromagnetic transducers.

resistive element represents mechanical losses within the piezoelectric plate and any acoustic or mechanical radiative energy. The electromechanical coupling coefficient may then be intuitively expressed as the ratio of the energy stored in the circuit associated with mechanical movement to the total energy stored in the circuit. Under linear DC conditions, this ratio η is given very simply by:

$$\eta = \frac{C_2}{C_1 + C_2} \tag{4.4}$$

Again using a qualitative argument that, like the qualitative arguments underlying Figure 4.3, may be confirmed analytically, we can obtain an equivalent to (4.4) in terms of the series and the parallel resonant frequencies of the circuit in Figure 4.3. The resonant frequencies are proportional to the square root of the capacitances and for reasonably high Q factors (greater than say 10) and for resonant frequencies that are not too far apart (50% in fractional bandwidth) (4.4) can also be expressed as:

$$\eta \approx \frac{f_s^2}{f_p^2} \tag{4.5}$$

where f_s and f_p are the series and parallel electrical resonant frequencies respectively.

This electromechanical conversion efficiency concept represents a *maximum* value for the transduction of input electrical energy into output mechanical energy. This is also, usually, simply a material property (the exception is the electromagnetic actuator). The overall efficiency depends critically on impedance matching, and here, again, much can be learned from a discussion of simple examples.

Amongst the simplest is the familiar hydraulic brake system, with or without servos, in any automobile. In this system the linkages enable the relatively low force:displacement (velocity) ratio of the driver's foot motion to be converted into very small (submillimeter) displacements of the brake pads but with orders of magnitude more force. This is of course a special case of a lever that is the most basic of mechanical impedance transformers.

The cantilever beam (Figure 4.4) is a slightly more complex lever that illustrates graphically how the input mechanical impedance may vary with position within a given structure. In qualitative terms a large force gives a small displacement at the root of the beam and vice versa at its tip. If we drive this beam with a dynamic force at a frequency well below resonance (that is, operating under stiffness control) by applying a force F at a distance x from the root of the beam, the resulting deflection at the point at which the force F is applied is

$$\delta_x = \frac{F \cdot x^3}{3EI} \tag{4.6}$$

The velocity at the point x is in quadrature with the displacement and is given by $\nu = j\omega\delta_x$. The mechanical impedance is totally compliant. Here E is the modulus of the material from which the beam is fabricated and I is the moment of inertia of the beam cross section, which for a rectangular section as indicated in the diagram is

$$I = \frac{at^3}{12} \tag{4.7}$$

The mechanical impedance at point x is

$$Z_{\text{mech}} = \frac{F}{j\omega\delta_x} = \frac{Eat^3}{j4\omega x^3} \tag{4.8}$$

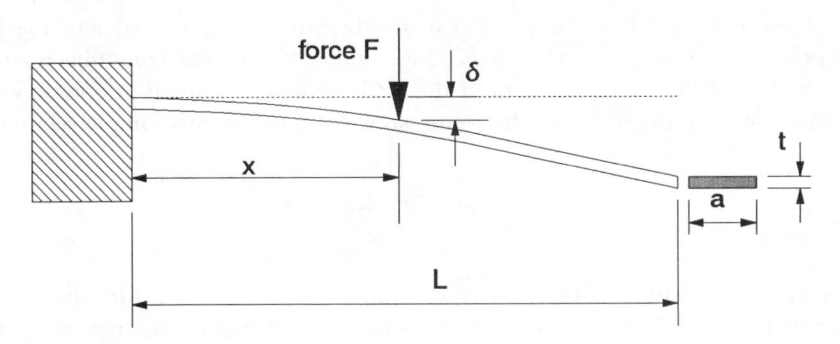

Figure 4.4 Geometry of a cantilever beam.

Equation 4.8 demonstrates the very wide range of potential mechanical impedances available along the length of a simple cantilever beam from the very high at the root of the beam to the modest at the tip. We also see the inverse frequency dependence that typifies stiffness control operation.

The impedances of actuator mechanisms can range from the very high typified by the piezoelectric device to the relatively low typified by many of the electromagnetic-to-mechanical conversion processes. The impedance and its dependence upon frequency for both mechanical source and mechanical load can vary over wide ranges and can be adjusted at the whim of the designer over a wide spectrum of values.

A simple example that illustrates the mechanical impedance matching concept—and one that is very important—concerns interfacing a piezoelectric actuator to, for example, a vibrating panel in order to perform active vibration damping. Under stiffness-controlled conditions (that is, well below resonance) the mechanical impedance of the source is given by

$$Z_{\mathrm{PZT}} = \frac{A_{\mathrm{PZT}} E_{\mathrm{PZT}}}{j\omega t_{\mathrm{PZT}}} \tag{4.9}$$

where A_{PZT} is the area of the transducer and t_{PZT} is its thickness.

All that is required then is to match the impedance of the PZT in (4.9) to that of the mechanical load, with due deference to the often subtle matching criteria discussed below. If we use the cantilever beam expressions as indicative of the behavior of the much less tractable vibrating panel, we find that the optimum position along the beam is obtained by equating (4.8) and (4.9):

$$x^3 = \frac{E_{\mathrm{beam}}}{E_{\mathrm{PZT}}} \cdot \frac{[at^3] \cdot t_{\mathrm{PZT}}}{4A_{\mathrm{PZT}}} \tag{4.10}$$

Examining this equation very rapidly demonstrates that the optimum position of a PZT drive on a cantilever beam is at a distance that is of the order of the beam thickness from the end of the beam. For most metals the Young's modulus is comparable to that of the PZT. The area of the PZT element is limited in one dimension by the value of a, the width of the beam, so we obtain the very simple relationship

$$x^3 = \frac{t^3 \cdot t_{\mathrm{PZT}}}{\ell_{\mathrm{PZT}}} \tag{4.11}$$

where ℓ_{PZT} is the length of the PZT element along the beam and is inevitably much larger than the thickness of the PZT element. The practical consequence of this is that all the experiments undertaken with vibration damping and cantilever beams place the actuator element as close as possible to the root of the beam. This is usually an intuitive reaction, but the reason for this lies within this impedance-matching concept.

Figure 4.5 is a plot of the mechanical impedance calculated for very low frequency operation (that is, for stiffness control) for an aluminum beam of length 1m and thickness 3 mm as a function of position. For low-frequency drive the *displacement* is in phase with the *driving force*, so the mechanical impedance is compliant (or in electrical terms capacitance-dominated). For comparison the mechanical impedance (or strictly the stiffness component thereof) of a PZT plate of length 1 cm, thickness 1 mm, and width equal to that of the aluminum bar is also given.

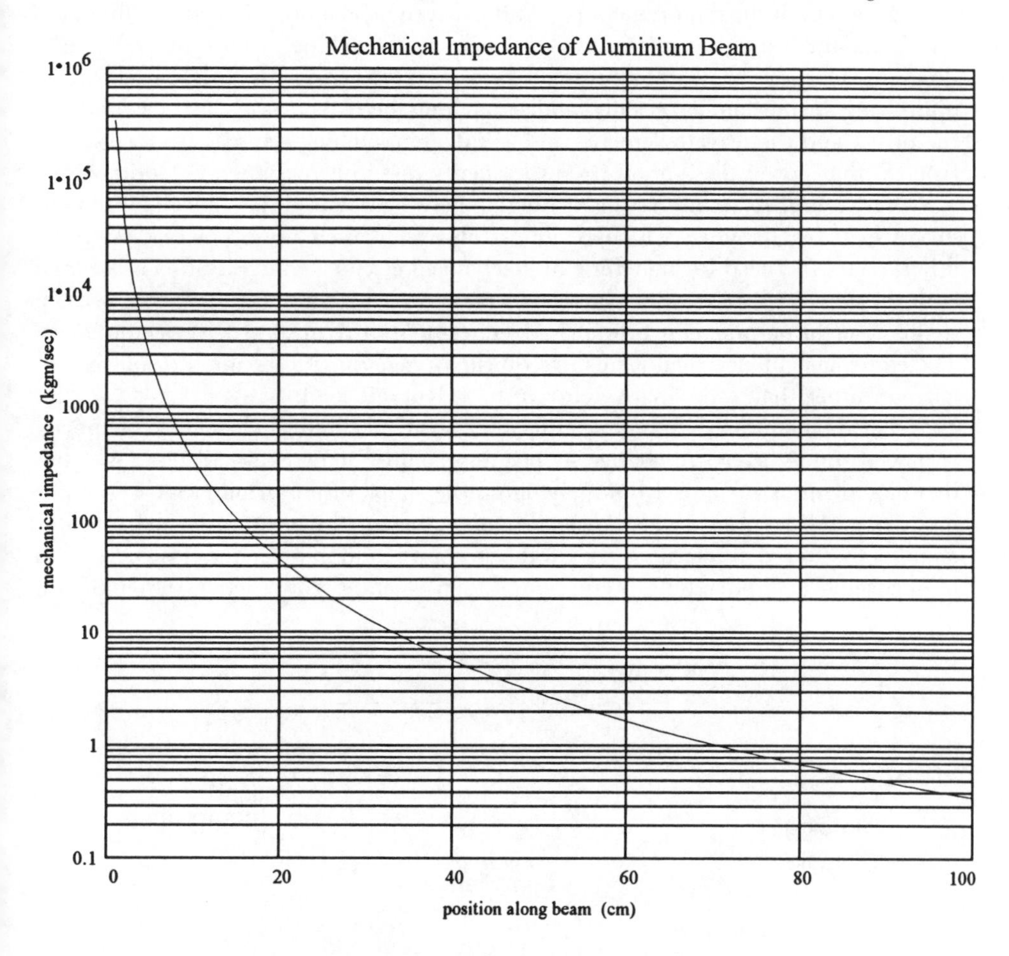

Figure 4.5 Variation of mechanical impedance with position along an aluminum beamwidth 100 mm, thickness 5 mm. In contrast, the mechanical impedance of a PZT element, width 100 mm, length 10 mm, thickness 1 mm is 100 M kgm/s.

The matching criteria must accommodate the type of drive that is required on the driven element. For example, if we wish to change the *position* of the end of the aluminum beam in a quasi-static fashion, the optimum course of action must be to maximize the *velocity* of the drive at the frequency at which the drive is applied. This in turn will maximize the available displacement: Maximum displacement and maximum velocity in a dynamic sense are equivalent ($v = j\omega x$). For stiffness control situations the equivalent circuit of drive and load is then as shown in Figure 4.6. In this situation the maximum *displacement* takes place when the driving and driven compliances are equal.

However, if the requirement is that the driven mechanical element (in this case, an aluminum beam) must do physical work, the matching criteria are inherently different. This is equivalent to putting a dash-pot at the end of the beam. In this case, optimizing the mechanical circuit entails matching the real parts of the impedances of the driver and the driven elements and conjugate matching the reactive terms. The result is that power dissipation from the energy reservoir may then be optimized.

This points to an interesting distinction between the traditional need for matching in electrical circuits, which is almost always focused upon power dissipation criteria, and the need for matching in mechanical circuits, which is often associated with the need to displace a mechanical element and then retain it in a new position. In this case, the parameter to be matched reduces to the driving and driven compliance. The equivalent in electrical terms is requiring a generator to create a displacement current rather than a dissipative current in a dynamic load.

For the majority of actuator applications, the mechanical element to be moved is indeed simply asked to change its position against some external force, whether this be moving a valve in a fluid, changing the shape of an airfoil cross section, or modifying the settings in an air-conditioning system. In most cases the actuator changes the stored potential energy in the structure, and in principle this energy can be recovered at a later date. Actuation is then essentially a reactive rather than a

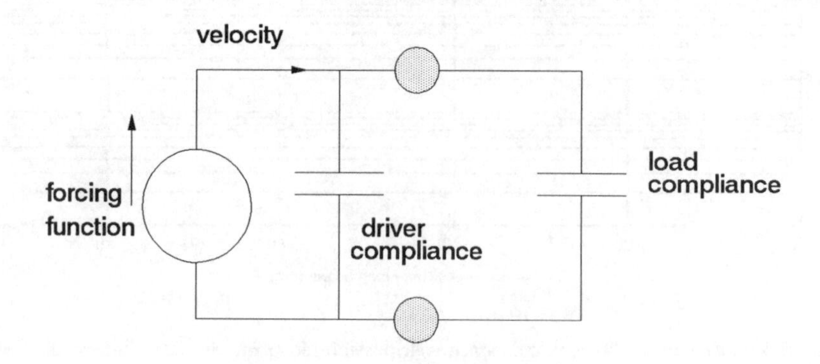

Figure 4.6 Generalized concept of impedance matching for mechanical circuits.

resistive process, though the speed with which the change can take place (reflected in the viscous elements within the mechanical circuit) is an important parameter, and through these viscous elements some energy dissipation does occur.

We can then, returning to Figure 4.1, look at the actuation process as an externally operated valve in which there are energy storage inefficiencies which preclude all the input energy appearing as useful mechanical energy followed by a driver into a mechanical load which must be in turn matched to the drive impedance. Much of actuator design entails the engineering compromise between these theoretical limits which determine the ultimate performance and associated practical issues concerning the size and cost of the entire actuation system.

4.3 ACTUATORS AND ACTUATOR MATERIALS—THE OPTIONS AND THE ISSUES

The majority of traditional actuator functions involve the application of relatively large forces over relatively large distances—typically tens of Newtons or in excess of tens of millimeters or more. This encompasses the vast majority of applications in land-based, marine, and aerospace vehicles and in highways and buildings. These forces and distances can usually only be generated through fluidic, electromagnetic, or internal combustion engine-generated drives.

The application of new approaches to the actuation problem must then seek to either:

- Improve the efficiency of traditional drive elements and their control systems;
- Improve the convenience in the control valve systems for traditional drivers;
- Respond to or define new applications that require entirely different driving functions.

Examples of the first of these include the use of new, high energy density magnetic materials to improve electromagnetic drive efficiency. The second of these may be exemplified by the frequent need for electromagnetically immune control inputs (for example, supplied optically) to operate large energy storage actuation systems, while examples of the third range from the need for precision actuators to line up optical circuits to the entire redesign of the actuation function to enable, for example, very slight changes in the shape of a wing section that may be realized piezoelectrically and, in turn, may completely modify the aerodynamics of the wing section. The latter is essentially another statement of the impedance-matching problem and may be viewed as applying a readily perturbed mechanical bias onto a specific mechanical function such that a small variation in this mechanical bias considerably changes the functionality of the system concerned. It seems intuitively obvious—though has yet to be confirmed by experience—that a wing section that can be modified using piezoelectric elements introducing very small perturbations must, by definition, be very close to a critical mode of operation. Military aircraft operators are familiar with this situation, but the frequent flyers of the world would probably prefer good old-fashioned flaps and rudders—at least for the time being.

The "smart materials" community has, for some time, focused its attention on actuator materials. The preceding discussion highlights the simple fact that most of these materials are unlikely to be able provide substantial actuation functions in their own right. They could perhaps become trimmers on the stiffness function of a structure, or they could perhaps operate directly in a control valve function. They may be able to introduce serious and significant changes in the performance of a mechanical structure provided this structure is biased close to a critical point that can be considerably modified using relatively high force:displacement product (that is, mechanical impedance) inputs.

In order to construct a discussion around actuator materials it is useful to reconfigure the problem into the form shown in Figure 4.7, which is a more complex version of Figure 4.1. This figure emphasizes the transfer of control source energy (reservoir B) into a controlled actuation source energy from reservoir A and the need for interface matching between the various elements within the actuator network.

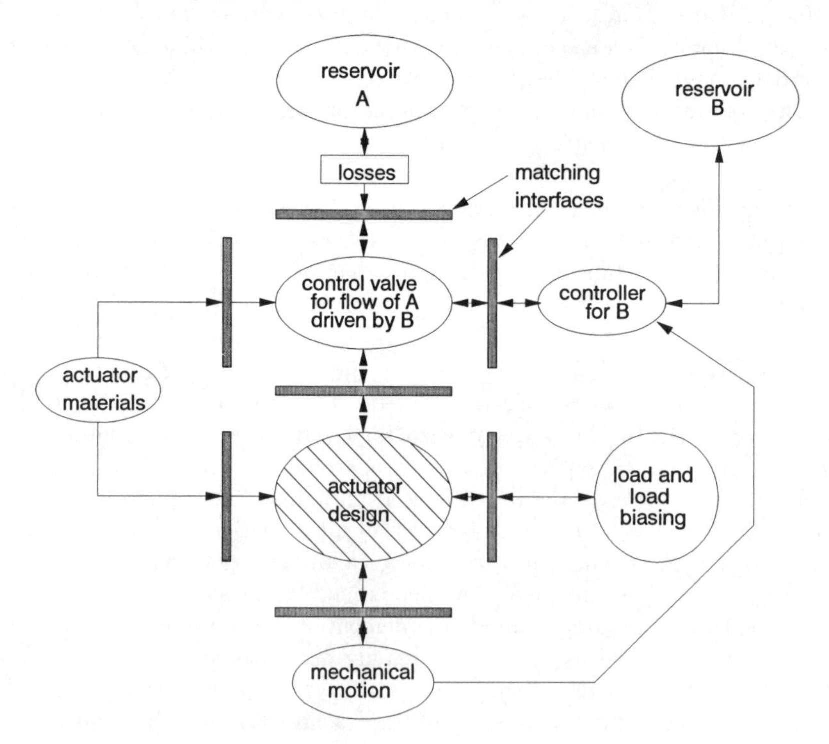

Figure 4.7 Matching criteria and conversion interfaces in actuator system design.

Additionally since the process is essentially reactive, the energy flows can be in either direction.

Table 4.1 summarizes the principal options for the actuation energy source, the control energy source, and the component materials that may contribute to the design of the controller and the final actuator. These materials will be discussed in more detail in the following sections. They are all, with the exception of electrorheological (and magnetorheological) fluids, forms of energy storage systems; and for all these systems the effective figure of merit must be the available force:distance product normalized to either material volume or weight. Electrorheological fluids are rather

Table 4.1
Elements of the Actuation Process

Primary Energy Sources (Reservoir A *of Figure 4.7)*

Energy Source	*Transmission Medium*	*Prime Mover*	*Comments*
Hydraulic	Pipework	Hydraulic ram	Invariably linear motion. Low loss transmission
Pneumatic	Pipework	Pneumatic ram or turbine	Linear or rotary. Low loss energy transmission
Hydrocarbon fuels	Pipework	Internal combustion engine, gas turbine	"Raw" energy converted on site. Usually rotary
Steam	Pipework	Steam turbine	Transmission difficult because high temperature. Rotary, very high powers
Electrical	Wiring	Electric motor	Easily controlled, high local efficiency can be lossy in transmission. EMI must be considered

Control Energy Sources (Reservoir B *of Figure 4.7)*

Source	*Comments*
Electrical	Most useful, virtually all actuators finally controlled by electrical energy. Simple to introduce precise feedback
Mechanical	Accelerator pedal best known! Usually involves a mechanical valve in a primary source transmission pipe; often this valve is electrically operated
Optical	Often useful as a control input to an (often optically energized) electrical control circuit for a mechanical valve assembly. Occasionally can introduce mechanical movement via thermal effects, but forces available are low
Fluidic	Fluidic amplifiers are occasionally useful—essentially hydraulic/pneumatic control of an output valve, but without moving parts (i.e., valve controlled by modifying fluid flow patterns). Require precision machining in complex shapes to retain flow profile

Table 4.1 (continued)

Control Valve Configurations	
Direct electrical drive	Used for all electromagnetic, piezoelectric, magnetostrictive, electrorheological actuator materials. Simple to control, simple to apply. Large controlled power level. Sometimes the material system is the ultimate primary actuator
Fluid drive control valves	Many actuators rely on on/off variable application of fluid pressure This is varied by modifying a flow constriction, usually itself electrically actuated and controlled. The electrically actuated valve should be impedance matched to the primary energy source
Actuator Materials and Technologies	
Piezoelectrics and electrostrictives	Typically stiff, small strain systems. Useful for moving loads over very short distances. Sometimes used with levers, e.g. "rainbow." Electric fields required are large
Magnetostrictives	Most usually used for hydrophone "pingers." Capable of larger strain than piezoelectric at similar stress levels. Need magnetic field
Electrorheological fluids	Electric field controlled viscosity. Used as liquid phase clutch plate. Physics of viscous forces implies essential loss. Particulates settle
Shape memory alloys	Material operates through transition temperature giving large strain (several %) with change in modulus. Thermally operated. Little strength in material
"Traditional" actuation	All the traditional hydraulic and pneumatic drive systems and linear and rotary electrical machines. Technology mature. Some improvements possible on weight of materials used
Microactuators	MEMS has recently come back to the fore. Micromotors can be configured as pumps or as rotary devices—the latter negligible torque but high speed

different. These are viscosity switches that are used to generate variable coupling forces between moving objects. These operations then drive switches (that is, a mechanical clutch) but with a difference. This switching process, since the force often relies upon the viscous drag in the medium, cannot be closed unless the drag is significant. Consequently, such a switch *must* be lossy. This is mitigated somewhat by the fact that these fluids can support a (usually modest) shear stress when electrically biased. Further, the driven section of the clutch must either be limited in torque transmission by the fluid characteristics or must move more slowly than the driver and accept the viscous losses.

With these general comments in mind we shall now examine some of the emerging actuator materials and endeavor to compare their properties and identify their potential application sectors. This applications discussion should also recognize the traditional and emerging roles of actuator systems. In particular, the traditional role has been to apply a relatively large force to a specific point to change the position of a load. The emerging role may well be to apply a similar force in a distributed fashion throughout the entire load, for example, in moving the shape of airfoil, or to modify the stiffness or mass loading of a structure in order to change its vibrational response. For the latter the actuator system becomes an integral part of the structure, while the former is a localized "bolt-on" load. This distinction is important since the presence of a distributed actuator inevitably modifies the normalized load-bearing capacities of the structure, including an increase in its weight through the presence of an actuator material that is usually not fundamentally load bearing. We shall explore these points in more detail in the following sections.

4.4 PIEZOELECTRIC AND ELECTROSTRICTIVE MATERIALS

4.4.1 Introductory Comments

Piezoelectric and electrostrictive materials are both ferroelectrics and as such have a number of features in common and exhibit many properties for which there are close parallels in the more familiar ferromagnetic materials. Of these the D-E loop is the most basic (Figure 4.8). This hysteresis loop characterizes the most basic ferroelectric materials into those with an open loop (the piezoelectrics) and those with a narrow loop (the electrostrictives). The area of the hysteresis loop represents losses due to ferroelectric domain wall movements, and the *coercive field* (E_c) gives an indication of the applied electric field required to create or to remove the permanent electric dipole moments within the material.

Using ferroelectrics as actuators implicitly exploits the application of relatively large electric fields and the translation of these fields into physical movements within the structure. The simple shape of the hysteresis loop immediately indicates a fundamental difference between piezoelectrics and electrostrictives—the former are inherently very significantly more lossy, and for both, the loss increases with frequency.

Ceramics play an important role in ferroelectric technology. The most important piezoelectrics (based upon PZT technologies) and the most important electrostrictives (PMN and its derivatives) are fabricated from pressed and fired powders and can therefore be made a wide variety of shapes and sizes. The permanent dipole moments in piezoelectrics are introduced by heating the material to beyond its Curie temperature, applying a large electric field, and cooling the structure to the operating temperature. This leaves a permanent charge density on the cooled material. No such processing (poling) is required for electrostrictives.

Ferroelectric polymers also exist—exhibiting either piezoelectric or electrostrictive properties—though only the former have received significant attention.

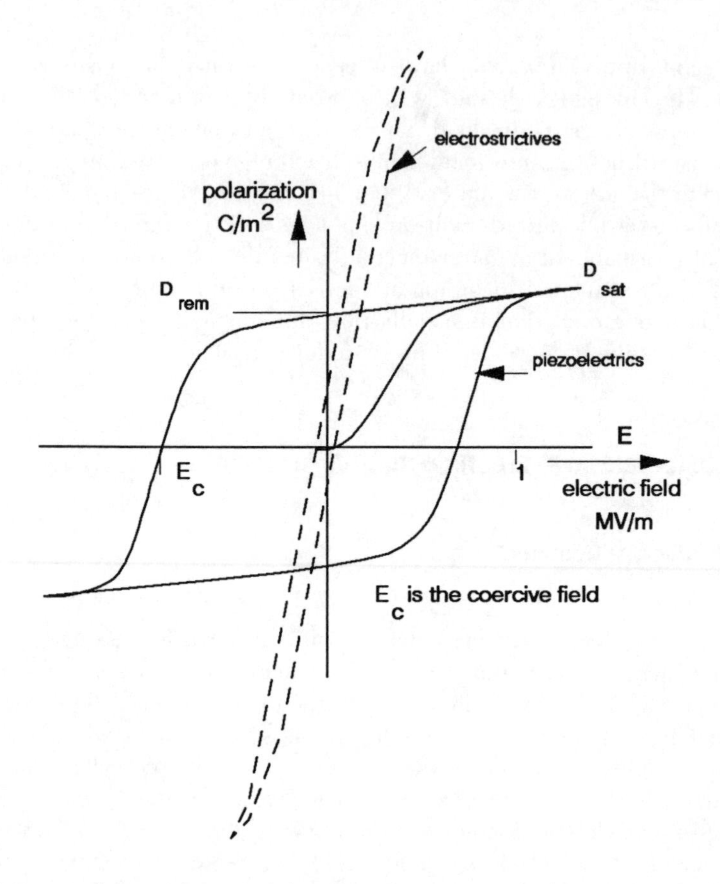

Figure 4.8 D-E curves for piezoelectric and electrostrictive actuator materials.

4.4.2 Piezoelectric Materials

Some detailed discussion of piezoelectrics has already been presented in Chapter 3, where their application as sensors was evaluated. When applied as actuators, the strains and the applied electric field are related by

$$\vec{S} = d\vec{E} \tag{4.12}$$

where d is the piezoelectric constant tensor defined in Chapter 3. There is, however, a major contrast between using piezoelectrics as sensors and as actuators. For the former application the strains involved create small electric fields—at most kilovolts/meter. For the latter the electric fields are very much larger and therefore exercise the piezoelectric material through a significant fraction of the D-E loop. In common with ferromagnetics, the result is that this process is both lossy and responsible for gradually decreasing the value of D_{rem}, inducing a gradual deterioration in performance.

Typical strain levels that can be applied by piezoelectrics are in the region of 0.1%, occurring at a field in the region of 1 MV/m. In general, this maximum field *must* be applied along the poling direction. The maximum field is limited to approximately 0.75 of the value of E_c.

A typical response for a commercial piezoelectric actuator system is shown in Figure 4.9 where the hysteresis is apparent. Excess drive levels produce progressive depoling and distortion of the hysteresis loop eventually into the butterfly shape shown in Figure 4.10.

Piezoceramics are also temperature-dependent and exhibit total depolarization above the Curie temperature, which, depending on the composition, ranges from 150°C to 400°C. As the Curie temperature is approached the value *d* decreases (Figure 4.11). However, the hysteresis in the basic characteristic implies that piezoelectric actuators are usually only useful when the positional feedback is applied. In the high operating temperature range, piezoceramic actuation systems are feasible provided that the displacements are compatible with the range of effective *d* values accessed as the temperature changes.

In addition to the gradual depoling introduced through exercising the material between extremes of applied electric field, there are other aging processes associated with the diffusion of dopants within the PZT. The deterioration associated with this is typically less than 10% per decade.

Piezoelectric ceramics have a Young's modulus that is typically 70 GPa at 0.1% maximum displacement. This gives a maximum energy storage capability within

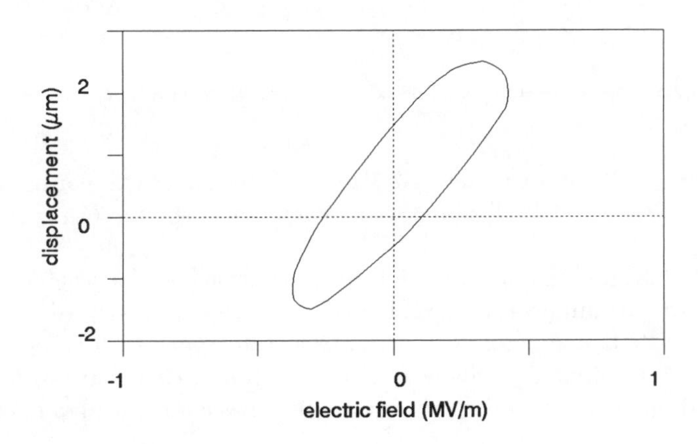

Figure 4.9 Typical characteristic for piezoelectric actuator.

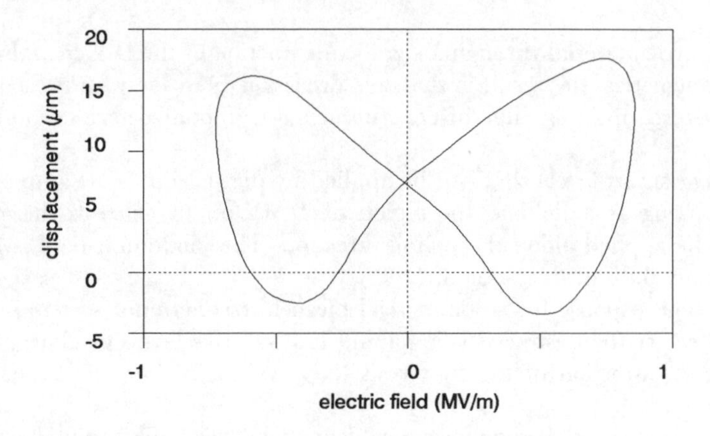

Figure 4.10 Typical response of depoled piezoelectric actuator.

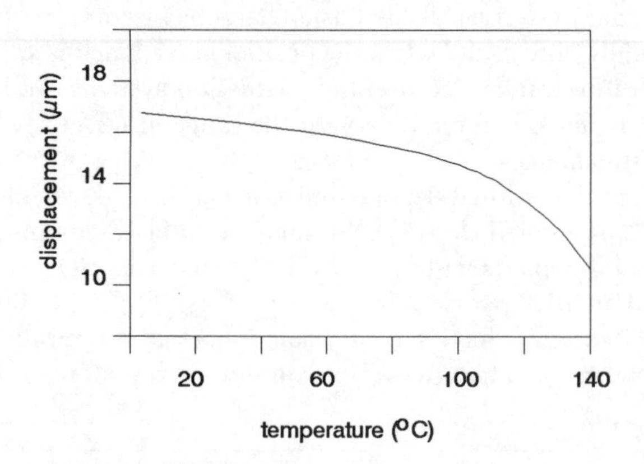

Figure 4.11 Typical temperature variation of displacement at constant drive for a piezoelectric actuator.

the piezoceramic of approximately 70 MJ/m^3. Hysteresis losses normally limit the operational frequency of PZT actuators to a few kilohertz due to the high electric fields involved.

Additionally, limiting the drive voltage to readily achievable levels of, say, 100V implies that the maximum usable PZT layer thickness is of the order of 100 μm. Some dynamic PZT modulators will perhaps operate under slightly higher voltages at larger thickness (often at mechanical resonance) but rarely involve structures more than 1-mm deep. The total displacement of the piezoceramic actuator δ is simply

$$\delta = d \cdot V \tag{4.13}$$

and is the order of 1 μm/kV. The value of V in (4.13) is obviously limited by the value of E_c, the coercive field, to avoid depoling. The term δ can be multiplied by the mechanical Q factor (which can have values up to approximately 1,000 but is typically roughly 100) but is still limited to a maximum strain of about 0.1%.

It is a rare actuator function that operates with a maximum of 1-μm displacement. Further the 1-μm displacement assumes that the actuator is not loaded. All actuator materials in practice work on a load line, as shown in Figure 4.12 for a typical piezoceramic. In order to increase the total displacement capability of a piezoceramic actuator, multilayer stacks (Figure 4.13) are often used. In this arrangement adjacent layers are in opposite directions and poled stacks to a maximum of 100 mm in height may be used to produce movements of 100 μm under negligible load or to move a stress in the region of 30 GPa through half this distance. 30 GPa corresponds to a load of 30k Newtons/mm^2. In practice strength constraints (600 MPa in compression, 80 MPa in extension) limit the achievable loads to of the order of tens of Newtons/mm^2. The operating regime is toward the bottom (low-stress region) of the load line curve.

Polymer dielectrics while very useful for sensing purposes have a Young's modulus of the order of 1 GPa and a maximum linear strain of about 700 μstrains. Energy storage capability is then only of the order of 1 MJ/m^3. This in turn implies a load-bearing capacity of a polymer of the order of 1 Newton/mm^2 in contrast with the previous much higher figures for ceramics.

4.4.3 Electrostrictive Materials

Electrostriction occurs in *any* material. When an electric field is applied the induced charges attract each other, thereby introducing a compressive force. This attraction

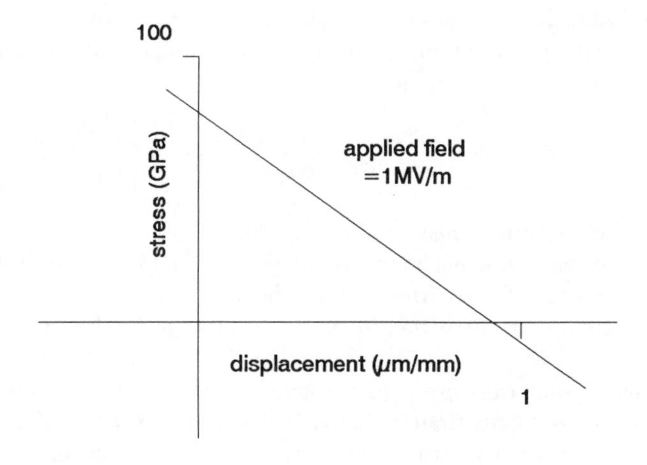

Figure 4.12 Load line for a piezoelectric material-in practice, compressive strength limits applied stresses.

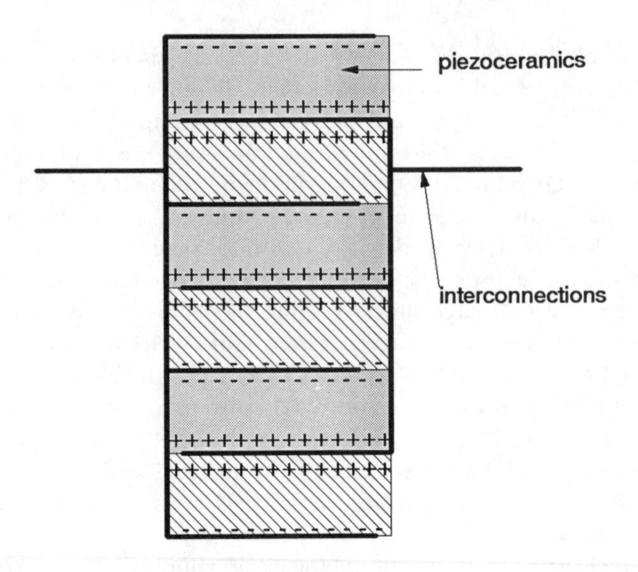

Figure 4.13 Piezoelectric actuator stack configuration: electrostrictives are similarly stacked, but poling is irrelevant.

is independent of the sign of the electric field and is easily shown to be closely approximated by a quadratic:

$$s = ME^2 = QP^2 \tag{4.14}$$

The strain s always lies along the axis of the electric field E, or, more correctly, along the axis of the induced polarization P. M and Q are the electrostrictive coefficients for electric field and polarization, respectively. The polarization is related to the electric fields according to the equation

$$P = \epsilon_0 \epsilon_r \hat{E} \tag{4.15}$$

and in general the dielectric constant ϵ_r is a tensor.

In most materials, the net effect predicted by (4.14) is relatively low. The exceptions are perovskite family (that is, ferroelectrics) for which the values of ϵ_r can exceed 20,000, giving strains of up to 0.1% for electric fields in the region of 1 MV/m.

The principal materials are lead manganese niobate:lead titanate (PMN:PT) and lead lanthanum zirconate titanate (PLZT). The former is particularly important.

A typical electrostrictive strain:field response curve is shown in Figure 4.14. This figure demonstrates that electrostrictives exhibit very low hysteresis (see

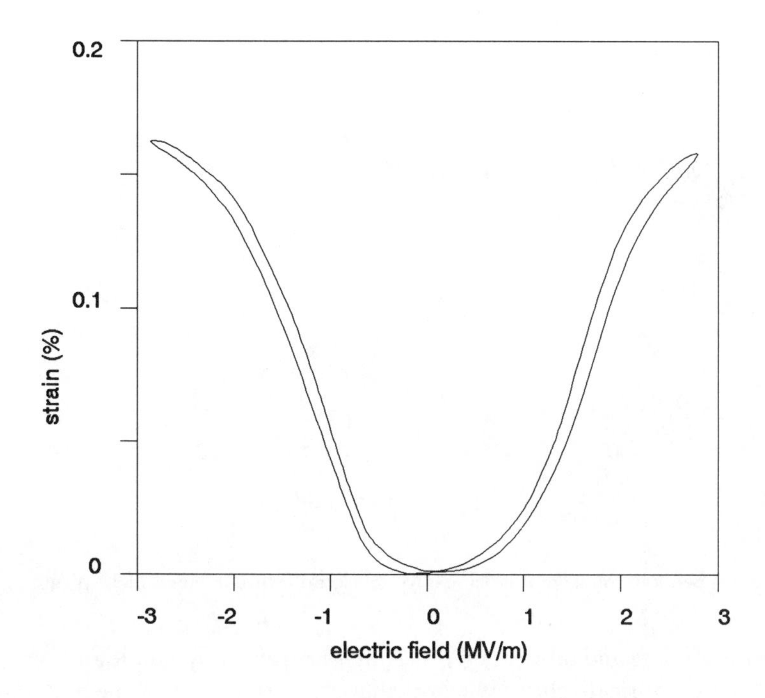

Figure 4.14 Typical response for electrostrictive actuator.

Figure 4.8) and, therefore, are low-loss devices. The figure also demonstrates that the quadratic approximation relating strain and electric field is only approximately true around the origin and should also recognize the influence of hysteresis.

These materials effectively rely upon a very high induced dielectric constant as an operating mechanism. Consequently they operate close to but slightly above the ferroelectric Curie point. Above this temperature there is—for pure materials—a $(T_A - T_c)^{-1}$ variation of dielectric constant on temperature. Consequently, the simplest form of electrostrictive material exhibits an interaction process that is very dependent upon temperature. However, the choice of the PMN hybrid and the addition of proprietary, usually rare earth, dopants to this material can produce a remarkably stable response. The general characteristics are shown in Figure 4.15. This shows relatively little thermal variation in the strain induced in the material for an applied field of 1 MV/cm over a range of –35 to +40°.

The maximum strain is again in the region of 1,000 μstrain and the modulus is similar to that of the piezoelectric, so the stored energy capabilities are very similar. However, the much lower hysteresis loss implies that these ceramics can be used to induce quite significant mechanical motion at frequencies up to 50 kHz, and new materials promise the possibility of operation at 10 MV/m with the commensurate increase in potential strain to approximately 0.25%.

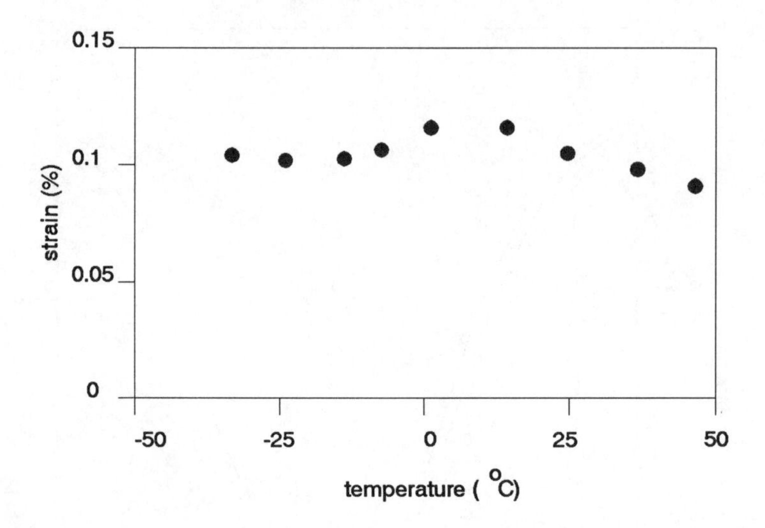

Figure 4.15 Temperature variation of strain in a typical electrostrictive at a bias of 1 MV/m.

The processing and fabrication procedures are relatively simple and very similar to those used to fabricate chip capacitors that are based on the same materials. The only disadvantage when compared to piezoelectric materials lies in the quadratic response. In all other features, electrostrictives are probably superior.

4.4.4 Piezoelectric and Electrostrictive Materials—Some Overall Comments

The materials have much in common. They can only induce very small strains at moduli in the order of 50 GPa to 100 GPa. They are ceramic materials and therefore are brittle and cannot by any contrivance be viewed as structural materials. They can, however, be configured through the firing process into a multiplicity of shapes and sizes and can also be realized in a multilayer form that enables the stacking of a large number of elementary thin sheets to reduce the necessary applied voltage. For both materials operation at approximately 1 MV/m is typical.

Both materials are relatively fast in their intrinsic response (potentially in the region of a few microseconds rise time). However, the high hysteresis in piezoceramics involves inevitable large losses that preclude their operation at high frequencies. Both materials in the "on" state consume negligible electrical power since the DC shunt resistance is extremely low (the hysteresis losses are due to domain wall movement).

These materials will be compared in a more global sense with other materials in the final section of this chapter. However, piezoelectrics and electrostrictives do have a wide range of applications ranging from the control of the hydraulic damper in the suspension system of the Toyota Lexus car to optical positioning equipment.

Both materials are high-modulus, low-displacement actuators limited by the need for leverage for the large-displacement actuation, the relatively low force:distance product and the intrinsic brittleness that implies any piezoceramic or electrostrictive actuator must be regarded as a structural load.

Finally, both phenomena do occur in a wide range of materials other than the commonly used ceramics. However, in the actuation context only the ceramics are useful. The polymers can be exercised through comparable strains but at very much lower material moduli. Polymer systems have, however, been used for demonstrations of complex material actuation structures where the flexibility and light weight of the polymer comes to the fore.

4.5 MAGNETOSTRICTIVE MATERIALS

Magnetostrictive materials shrink in the presence of a magnetic field. The basic phenomenon is very similar to electrostriction. Magnetostrictive materials are ferromagnetic with extremely narrow B-H loops. They comprise alloys of iron, nickel, and cobalt doped with rare earths. The most effective magnetostrictive material is probably TERFENOL-D, an alloy of terbium, dysprosium, and iron $Tb_xDy_{1-x}Fe_y$ with $0.27 \le x \le 0.3$ and $1.9 \le y \le 1.95$. The name combined TERbium, FE, and the Naval Ordinance Laboratory (now the Navel Surface Weapons Center) where the alloy was invented. Terbium is arguably the rarest of rare earths and so TERFENOL is also the most expensive of magnetostrictive materials.

Not surprisingly the phenomenon is at its best in crystalline material, hence magnetostrictives are typically seeded from the melt on to a crystalline rod. This zone-refining process limits the size and geometries available into wires that, for TERFONOL-D, are rarely in excess of 20 cm in length. Even the more common magnetostrictives (magneks, permedyn) remain expensive and restricted in geometries to prisms of a few tens of centimeters long and a centimeter in diameter. Additionally they are difficult to machine without impairing the magnetostrictive properties.

The principal advantage of magnetostrictives are that they are capable of exercising strain levels up to 2,000 μstrains at moduli typically in the order of 100 GPa. The additional fact that they can be fabricated in lengths of the order of 200 mm implies that quite large movements are available. Further, in common with their electrostrictive cousins, they have narrow hysteresis loops (a very few percent) and therefore are relatively low-loss materials. Typical characteristics are shown in Figure 4.16. This characteristic can also be modified by prestressing the sample and to first order the behavior is quadratic. The forces that can be generated are very large; for example, a 10-mm diameter element held to a 0.1% strain generates forces of the order of a kiloNewtons depending upon the detailed structure.

For quasi-linear operation, the magnetic bias is obviously required and is typically of the order of 100 kA/m. The necessary driving fields can extend up to several

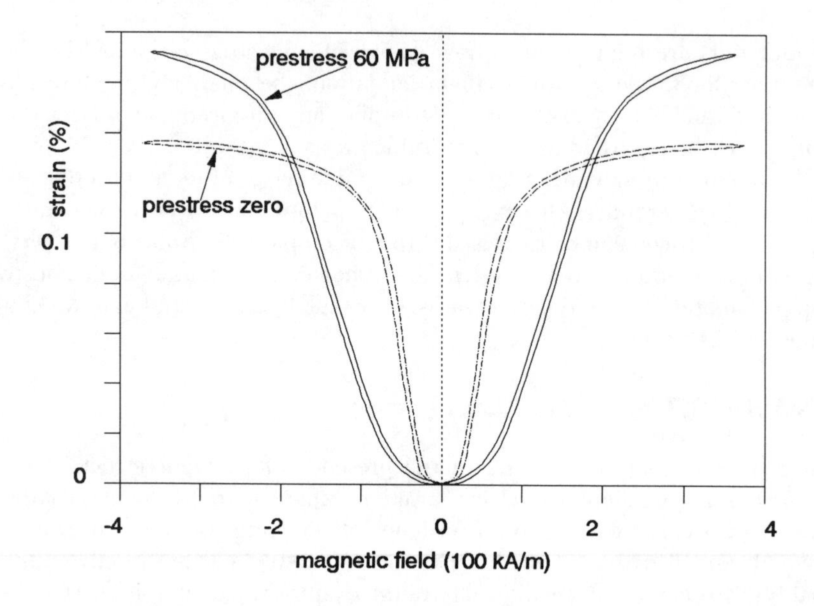

Figure 4.16 Strain response of a typical magnetostrictive actuator, showing the effect of prestress on response.

tens of kiloAmperes/meter, so the design of drive coils is critical and, even though the materials themselves are relatively loss free, Ohmic losses in the coil can be significant. The materials in common with electrostrictives tend to be brittle, so the prestressing process (which forces the molecules a little closer together and effectively increases the permeability) serves the dual process of avoiding excessive tensile strengths that may compromise the material's integrity. The low hysteresis, the equivalent excellent repeatability of positional tolerances, and the relatively large element geometry are significant advantages of magnetostrictives. They are, however, relatively rarely used-their most common application is as "pingers" for sonar systems. Their extremely high energy storage capability is somewhat offset by the cumbersome but necessary driving coil arrangements and the inherent expense of the material itself.

4.6 SHAPE MEMORY ALLOYS

Shape memory alloys, which usually comprise a mixture of nickel and titanium, have been known for about a quarter of a century. Their metallurgical features give them rather amazing properties. When cooled to below a critical temperature their crystal structure enters the so-called martensitic phase. In this state the alloy is plastic and can easily be manipulated through very large strain ranges with relatively little change in the material stress.

When the alloy is heated above the critical temperature (which can be varied by varying the composition of the alloy) the phase changes to the so-called austenitic phase. Here the alloy resumes the shape that it formally had at the higher temperature, in particular has high strength and high modulus, and behaves very much like a normal metal. Additionally the alloy shrinks by typically several percent in transferring from the low-temperature martensitic phase to the high-temperature austenitic phase. Figure 4.17 shows the stress-strain curve for nitonol—one of the more commonly used shape memory alloy systems.

The use of shape memory alloys as actuators relies upon operating them normally in the low-temperature plastic martensitic phase and constraining them within some structural assembly in this state. When the alloy is heated through to the austenitic phase, the external structure constrains the alloy from returning to its remembered shape. Consequently, significant stresses (several hundred MPa) are generated within the alloy and these, in turn, can stress the structure within which the alloy is mounted.

There are other changes in materials properties that occur when the alloy goes through the transition temperature. Of these the most important, in addition to the dramatic change in modulus, is the parallel change in material mechanical losses. When in the low-temperature martensitic state the plastic material is also very lossy, and therefore the shape memory alloy can be used as a variable damping material.

The properties of shape memory alloys are very complex and depend critically upon the composition and past history. In particular, the greater the magnitude of

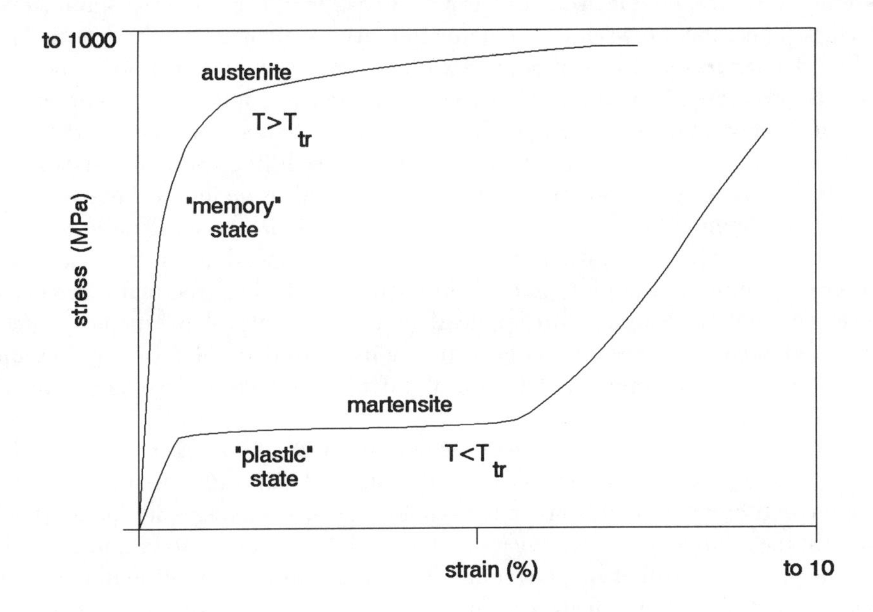

Figure 4.17 Stress-strain curves for a shape memory alloy in each of its two states.

shape recovery during the phase change, the shorter the fatigue life of the material (see Table 4.2). Consequently the actuator function is probably limited to rather less-traumatic strain changes than, for example, the pipe connector, which is used once. The transition temperatures of the various alloy formulations can also vary over a range from –150° to +150° with the high strain change materials tending to have lower transition temperatures; for example, the pipe connectors are often applied after cooling in liquid nitrogen.

The remarkable physical properties of a typical shape memory alloy-nitonol-are summarized in Table 4.3. From this we can see that SMAs are relatively heavy, are reasonably strong, and can withstand high strain levels. In operation, they must be forced through the critical temperature, obviously by heating or cooling. The 10° to 30° hysteresis in the transition temperature must be accommodated into the heating and cooling cycle.

This need for thermal drive is probably the greatest single operational difficulty with shape memory alloy actuator systems. It fundamentally limits both the available efficiency of the system and the speed of response. The former may be viewed as a classical heat engine, so basic thermodynamics indicates that the *maximum* achievable efficiency must be of the order $\Delta T/T$. A usable temperature switch is typically a few tens of degrees centigrade, so achievable efficiencies of the order of 10% are typical.

Response times present an even greater paradox. To minimize the power consumption of the SMA in the austenitic phase, the actuator rod should be insulated as effectively as possible from its surroundings. In the limit, this minimizes the mechanical contact between the actuator rod and its surroundings and so, since the SMA becomes less and less in contact with the load, could constrain its actuation role. However, good insulation (that is, a large thermal resistance between the actuator and the surroundings) in turn implies long thermal time constants and leisurely reaction times. Achieving rapid responses implies reducing thermal resistance and therefore increasing the power consumption required to maintain the actuator rod in the high-temperature state. If the ausenitic state is in thermal equilibrium with its surroundings (for example, in the pipe connector application), then this additional power consumption is not a problem—and neither is the long thermal time constant since these are typically one-off applications for which the shape memory effect is only used once. However, for actuator functions these thermal drive considerations present a serious limitation on the applicability of the materials, frequently imposing the need for forced cooling.

The greatest single advantage of shape memory alloys lies in their very large energy storage capacity, which is explored in more detail in the final section of this chapter and compared with other materials. The disadvantages lie in the thermal drive, in the relatively low efficiencies and in their susceptibility to fatigue. Additionally the precise control of shape memory actuators is rather complicated due to their inherent highly nonlinear features. The shape memory alloy has, however, made its mark as the ultimate pipe connector, as a relatively strong damping material, as the

Table 4.2
Properties of Some Shape Memory Alloys

Alloy	*Structure of Low-Temp. Phase*	*Thermal Hysteresis (k)*	*Shape Recovery Strain (%)*	*Fatigue Life (cycles)*	*Application*	
Ni-Ti Ni-Ti-Fe	Rhombohedral	2~3	1	1,000,000	Two way (repetitive)	Actuator (long life)
Ni-Ti-Cu	Orthorhombic	10~15	4~5	10,000		Actuator (large stroke)
Ni-Ti Ni-Ti-Fe	Monoclinic	20~40	6~8	<100	One way (static)	Connector Coupling
Ni-Ti-Nb	Monoclinic	60~100	6~8	—		

Table 4.3
Typical Properties of Shape Memory Alloy Materials

Property	*Unit*	*Ni-Ti*	*Cu-Zn-Al*	*Cu-Al-Ni*
Max. strain (one-way effect)	%	6 to 8	4	5
Max. strain (two-way effect)	%	4 to 5	1 to 2	1.2
Shape memory strain	%	6 ($N = 100$)		
Super elastic strain		2 ($N = 10^5$)		
		0.5 ($N = 10^7$)		
Effect stability	cycles	100,000	100,000	1,000
Recovery stress	MPa	400 to 700		
Tensile strength	MPa	800 to 1000	400 to 700	700 to 800
Elongation to break	%	40 to 50	10 to 15	5 to 6
Young's modulus	GPa	20 (martensite)		
		80 (austenite)		
Poisson's ratio		0.33		
Density	g/cm^3	6.5	8	7
Transformation temperature (depending on composition)	°C	–200 to +100		
Max. As temperature	°C	120	120	170
Max. operating temperature	°C	400	160	300
Hysteresis	°C	10 to 30	10	35
Efficiency	%	10		

N is number of cycles.

drive in thermally activated switches (those whose electric kettles operate with SMA cut-outs will recognize the hysteresis) and for some forms of specialist electromechanical actuation systems where the flexibility of linear or rotary design and high energy density offset the low efficiencies and long time constants.

4.7 ELECTRORHEOLOGICAL FLUIDS

Electrorheological fluids have been known for over a century, and their potential in engineering systems was probably first recognized in a patent application for an electrically operated clutch dating from the late 1940s.

An electrorheological fluid is one whose viscous properties may be modified by applying an electric field. The prototype electrorheological fluid is a mixture of silicone oil and corn starch, but a wide variety of mixtures of nonconducting solids and colloidal suspensions show electrorheological properties. The basic operating mechanism is very simple. In the neutral state the particles are uniformly distributed throughout the fluid. When the electric field is applied, the relatively high dielectric constant of the particles captures this electric field, redistributes the charge densities within the particles, and causes them to stick together in fibrils. The presence of these fibrils considerably modifies the viscous properties of the fluid. Figure 4.18 shows this process in conceptual form and illustrates intuitively how the presence of these fibrils could considerably modify the viscous properties of the fluid as viewed by measuring shear forces between the two electrically charged forces.

Electrorheological fluids are non-Newtonian, which simply means that the viscous behavior as portrayed in a graph as shear stress versus shear strain rate is nonlinear and typically contains an offset term that allows the fluid to support some shear stress in the absence of shear strain rate. Typical viscosity and shear stress (in the absence of any shear strain rate) curves for an electrorheological fluid, shown in Figure 4.19, demonstrate considerable changes in viscous properties are only achieved at relatively high fields in the order of 1 kV/mm.

These curves demonstrate the basic properties of electrorheological fluids that may be exploited in practice:

- They are only useful to transfer *shear* stresses.
- These shear stresses are limited by the offset shear stress (the right-hand axis of Figure 4.19) unless slippage between the input and output planes can be tolerated.
- The viscosity of these fluids (that is, their responses to shear strain rates) can be increased significantly by applying an electric field. This increased viscosity increases the *losses* in the fluid introduced as a result of the shear strain rate. Consequently application of an electric field to an electrorheological fluid can in turn introduce a variable damping coefficient.

There are a number of operational features that quickly emerge from this discussion. In particular, we have the following:

- The finite conductivity of the fluid implies a power dissipation even in the absence of an external shear rate.

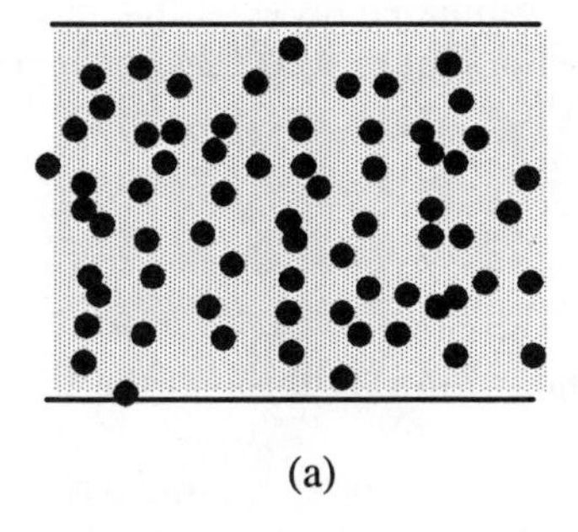

(a)

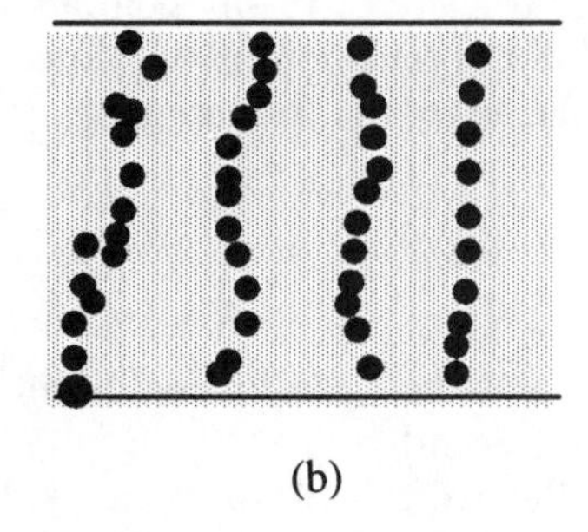

(b)

Figure 4.18 Electrorheological fluids: the generation of fibrils in the presence of an electric field. (a) No field applied; (b) field at about 1 MV/m.

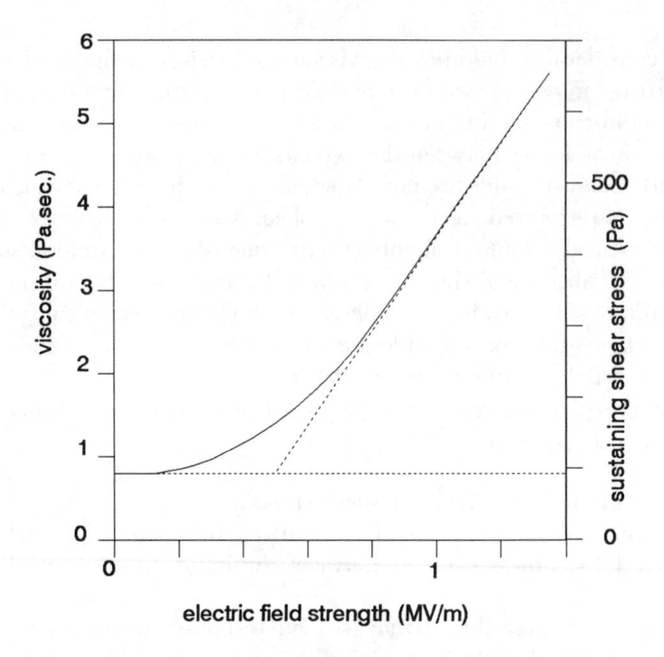

Figure 4.19 Properties of typical electrorheological fluids.

- In the presence of a shear rate, the power dissipation increases yet further.
- The operation of the fluid depends critically upon the homogeneity of the colloidal suspension. Sedimentation is therefore a considerable problem and, even though in principle the particle density can be matched to the fluid density, this can only apply over a relatively limited temperature range.

As with other actuator materials, the parameter of interest has to be a parallel energy transfer or dissipation factor. To obtain an estimated order of magnitude it is helpful to consider the concentric structure shown in Figure 4.20. The shear stress transferred by the fluid may be approximated by

$$\sigma = \sigma_{\text{strat}} + \eta \frac{dv}{dr} \tag{4.16}$$

where η is the viscosity coefficient and dv/dr is the strain rate through the fluid in inverse seconds.

Referring back to Figure 4.20 and assuming that r_1 is approximately r_2 we obtain a power transfer coefficient (that is, torque × angular velocity) of

$$P = \sigma r_1^2 \ell \omega \tag{4.17}$$

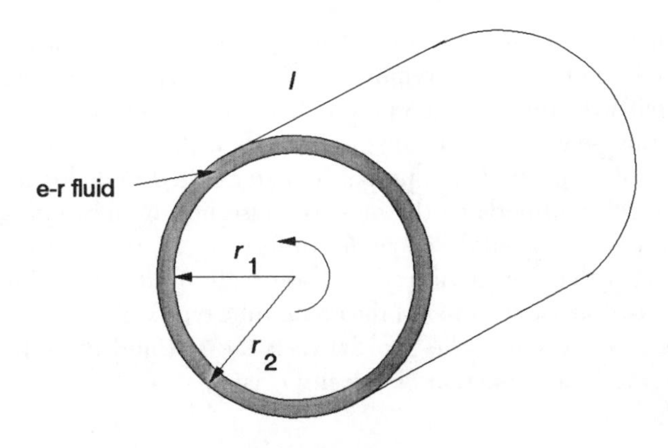

Figure 4.20 Representation of a potential e-r fluid clutch.

Figure 4.21 is a plot of the power-transferring capability for such a clutch (assuming zero slippage, that is, zero strain rate) for an *rl* product of 1 m^2. at a typical automobile engine speed of 3,000 rev/min (taken as 300 rads/s). This indicates that useful power—up to 50 kW—can be transferred through a 1 m^2 *rl* product with $r_1 = 0.25$m. However, in the automobile context simple friction clutches transfer more power

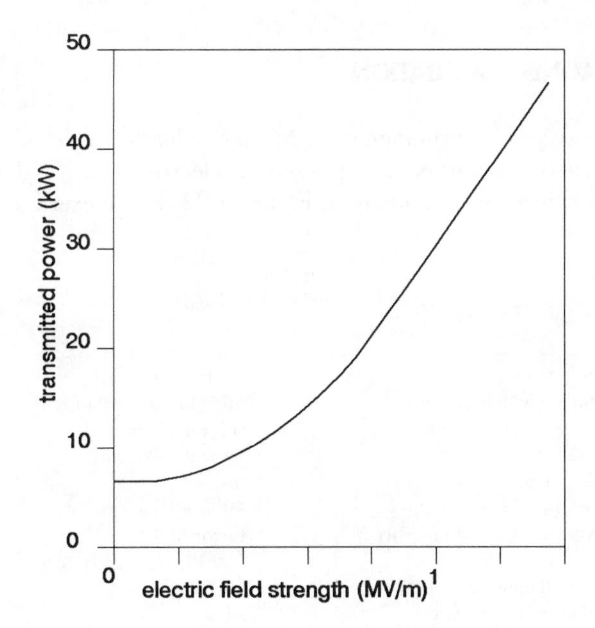

Figure 4.21 Nonslip power transfer characteristics of a simple e-r clutch with $r_1 l = 1$ m^2 and $r_1 = 0.25$m.

through less than 5% of this total contact area and also have the advantage that they can be switched totally "off" by removing the friction contact. Even at zero electric field, this clutch will transfer some power. If the driven element is prevented from moving, then this power has to be dissipated within the clutch itself.

Electrorheological fluids are probably better suited to applications other than drive train control, and perhaps the most promise lies in vibration damping where the switchable viscosity coefficients comes to fore.

The changes that can be introduced are really quite large. For example, we take a hypothetical shock absorber of the rod in tube type where the rod is of diameter d, the tube is stationery and the gap between the rod and the tube is g. We can estimate the force F as a function of velocity v as:

$$F \approx \eta \frac{v}{g} \pi d \ell \tag{4.18}$$

where ℓ is the interaction length. For a diameter of 5cm, a gap of 1mm and an interaction length of 1m, a typical electroheological fluid will enable the viscous drag on the rod to switch from 150 Newtons to about 1000 Newtons at a velocity of 1m/second.

The general properties of typical electrorheological fluids are summarized in Table 4.4.

4.8 ELECTROMAGNETIC ACTUATION

Large cranes have used electromagnets to lift heavy loads for decades.

Electromagnetic actuators (as opposed to electric motors) share a common geometry that is shown schematically in Figure 4.22. Force exerted on the load by

Table 4.4
Typical Properties of ER Fluids

Particle size range	1μ to 100μ
Suspended particle content	15% to 40% volume fraction
Voltage requirement	several kV/mm
Power density	several mW/cm^2
Response time	ms
Viscosity (at E = O)	1000 mPa.s. (at 25°C)
Maximum increase of viscosity	factor of 10
Shear stress	500 Pa (at 1 kV/mm)
Operating temperature	–30°C to 300°C
Fluid dielectric constant	~1.5
Particle dielectric constant	~10

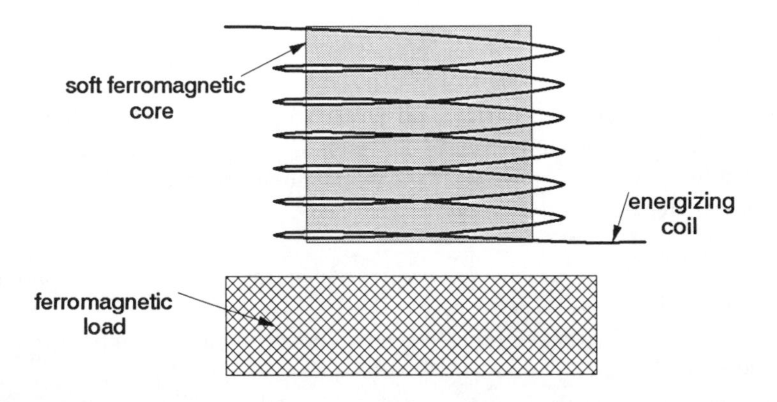

Figure 4.22 Electromagnets.

the magnetic field generated in the core is exactly analogous to the electrostatic force and is given by

$$\frac{F}{A} = \frac{B^2}{2\mu_o} = \frac{\mu_0 \mu_r^2 H^2}{2} \tag{4.19}$$

This equation highlights the importance of the effective relative permeabilities of the core and the load—expressed here as μ_r^2—in determining the lifting power. For a relatively cooperative load, an effective value of B can be a significant fraction of 1 Tesla; so the force per unit area exerted by the electromagnetic actuator is of the order

$$\frac{F}{A} \approx \frac{0.5 \times 0.5}{2 \times 4\pi \times 10^{-7}} = 0.1 \text{ MPa} \tag{4.20}$$

This force may be exerted over a distance of several centimeters and is equivalent to a load-carrying capacity of 10 tons/m^2. With high permeability cores the value of B can be driven much higher and lifting capacities for electromagnets of up to 100 tons/m^2 are achievable. When we reduce this to the scale of the conventionally viewed actuator system whose dimensions may be of the order of centimeters, we still find that a core with an area of 10 cm^2 is capable of supporting loads in the order of 100 kg. Finally even though the stresses created by the electromagnet are one, perhaps two, orders of magnitudes less than those created using, for example, a piezoelectric or electrostrictive actuator, the distances over which this force can operate are perhaps two orders of magnitude higher. The energy-changing capacity of the two media is then broadly comparable. Where the electromagnet scores is that it can be made really large—hence it becomes capable of these substantial lifting capacities.

4.9 HYDRAULICS

Hydraulics are mentioned primarily as an energy *transfer* medium. Hydraulic systems are not actuators in their own right: they must be driven by a pump or pressure tank—usually the former. However, hydraulic systems offer a number of attractions as energy transfer media:

- They are very low loss since relatively little energy is stored or dissipated within the hydraulic fluid.
- They can transfer very significant stresses operating at tens to hundreds of MPa (100 to 1,000 Bar).
- They consume very little energy except when in use. The losses arise from the viscosity of the fluid, which is designed in the fluid characteristics to be very small.

Traditionally hydraulics have been used in piston and cylinder actuator arrangements, and such arrangements remain very common use—the most familiar being the automobile brake cylinder. There have been a few, as yet tentative, attempts at distributed hydraulics incorporating capillary hydraulic feeds into structures in order to control effective stiffnesses. This is a common biological actuation system used in plants and animals, but since engineering structures tend to be orders of magnitude stiffer than biological structures, the effective changes in the usable stiffness that can be introduced by hydraulic actuators are much smaller for the engineer than they are for nature's creatures. Even with hydraulics operating at 100 MPa the changes in a structure whose natural stiffness is 50 GPa are small indeed.

4.10 CONCLUDING COMMENTS—THE ROLE OF ACTUATORS AND ACTUATOR MATERIALS

For some reason actuator materials are the ones that the populists love to refer to as "smart." My personal view is that nothing could be further from the reasonable truth. Actuator materials turn one or more (and usually more—they are almost all temperature-dependent) input stimuli into an output stimulus, which in our restricted definition is a change in mechanical energy. This mechanical energy is either stored within the actuator material itself or transferred into an external load depending on the application. For all actuator materials the energy transfer capability is a useful figure of merit. In Table 4.5 we present a summary of the relative properties of the various actuator materials that have been discussed in this chapter. The surprising feature of this table is that in terms of energy transfer capability, various materials are broadly comparable with possibly the shape memory alloy outshining the others—but only on a normalized volume basis. The fact that relatively small cross-sectional

Table 4.5
Comparison of Actuator Materials and Techniques

Material/Technique	*Maximum Displacement*	*Stress Capability*	*Energy Change ($J.m^{-3}$)*	*Dissipation "on" or "off"*	*Comments*
Piezoelectrics	1000 $\mu\epsilon$ at 1 MV/m	50 GPa to 100 GPa (Ceramic) 1 GPa (Polymer)	~100 MJm^{-3} (ceramic) ~1 MJm^{-3} (polymer)	Very low in both states—requires holding field	Open D-E loop implies high hysteresis. Linear field/strain characteristic. Ceramics can be configured in many shapes. Multilayer stack for large displacements. Strength ~100 MPa
Electrostrictives	1500 $\mu\epsilon$ at 1 MV/m 2500 $\mu\epsilon$ anticipated at 5 MV/m to 10 MV/m	50 Gpa to 100 GPa	~100 MJm^{-3} to 150 MJm^{-3}	Very low — excellent dielectric as piezoelectrics	Closed D-E loop gives low hysteresis and so can operate at up to 100 kHz. Parabolic field/strain characteristic. Ceramics fabricated in variety of formats. Strength as piezoelectric
Magnetostrictives	2000 $\mu\epsilon$ at 200 kA/m	100 GPa	to 200 MJm^{-3}	Little material dissipation but coil currents can be significant	Long (20 cm) prisms of 1 cm^{-2} give large actuator subunits. Refining processes and materials costs give very expensive basic material. Machining after fabrication difficult. Best preloaded. Quadratic current/ displacement response. Strength to 1 GPa
Shape memory alloys	to 5%	50 GPa	to 1 GJm^{-3} (estimate difficult because of gross changes in material properties)	High "on" dissipation due to need to heat above ambient unless used as one off "clamp," e.g., pipe connector	Large energy change is attractive but only available in wires to few millimeters in diameter, and need for thermal drive implies slow response. Switchable cooling necessary for rapid "off" transition. Most effective for stiffness modulation, alternatively, load against spring to activate low-temperature phase

Table 4.5 (continued)

Material/Technique	*Maximum Displacement*	*Stress Capability*	*Energy Change* ($J.m^{-3}$)	*Dissipation "on" or "off"*	*Comments*
Electrorheological Fluid	N/A	Maximum shear stress ~500 Pa at 1 MV/m	Does not store or release *usable* energy	Several mW/cm^{-2}. Higher when flowing and for thicker (>0.5 mm) layers	Field controlled viscosity useful for damping control Electrically switchable power train feasible but limited by shear stress-bearing capacity before inducing flow and significant mechanical losses
Electromagnetic Actuators	Centimeters: depends on pole area	~1 MPa	Assuming a 0.1-m working distance implies ≈ 100 kJ/m^2	High dissipation in "on" coils unless super-conducting	Simple on/off control but relatively limited functions—usually only lift:not-lift states available (except linear and rotary machines)
Hydraulics	Can be very large >0.1m	to 1000 Bar (100 MPa)	~1 MJ/m^2	Low loss no dissipation in on/off states. Needs holding force	Extremely versatile but use currently largely limited to piston in cylinder systems

areas of shape memory alloys are all that is available tempers their potential somewhat. Their need for large driving currents and the fact that they operate on thermal transition phenomena also makes them somewhat inconvenient to use.

The principal technical issue then is dominated by a combination of the total forces required and the mechanical impedances for which these forces should operate. Piezoelectrics electrostrictives and magnetostrictives tend toward the stiff, small displacement system, while hydraulics and electromagnetics are immediately into longer displacements but at lower forces per unit area.

Another important, but often not overriding, issue lies in the energy consumption of the actuation system. Ideally (Figure 4.23) the actuation system should be as near as possible to 100% efficient as defined by the total change in mechanical energy in the system to be moved compared to the total energy consumption throughout the life of the actuation system. In Figure 4.23 we attempt to compare, at least qualitatively, the behaviors of the various material systems, and it is here that the benefits of low-loss actuation, particularly based on piezoelectric and electrostrictive medium and on hydraulics, become apparent. All the other media require significant latching power to retain the mechanical system in position.

Finally, and to return to the concept of "smart," it is worth re-emphasizing that these materials are pathologically dumb. The smartness appears when the actuator

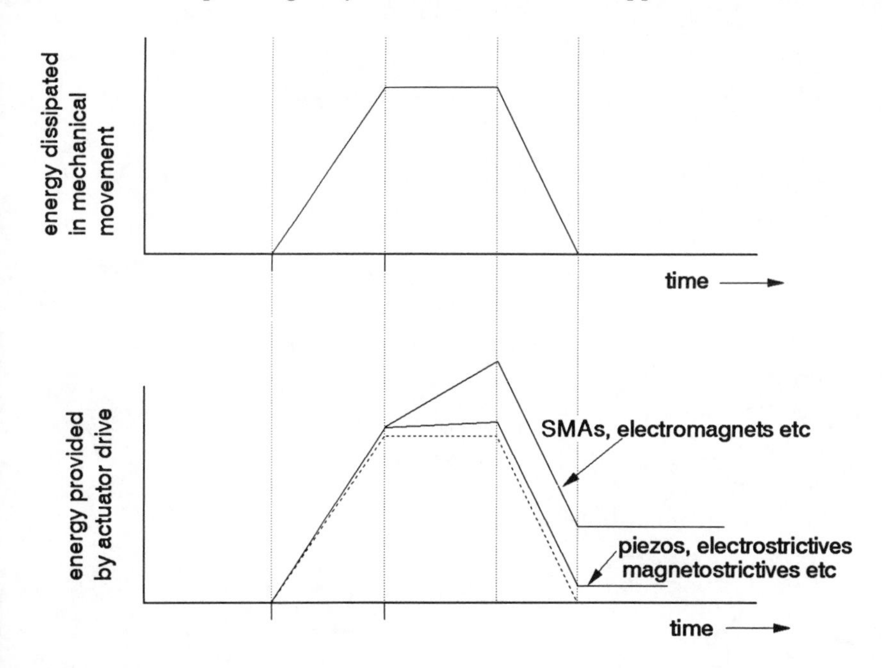

Figure 4.23 Energy dissipation in the actuation process. Does the actuator need to dissipate energy when the load is stationary?

material is combined with a sensory material into a material system, perhaps als
incorporating electronics, which is capable of *selecting* from a range of alternativ
responses depending upon the diversity of sensory inputs. At present this "intelli
gence" does exist in man-made systems using electronic processing, but the syntheti
biological equivalent remains illusive. Some of these issues are considered in mor
detail in Chapter 6 when we consider some of the applications of smart structure
and materials.

Chapter 5

Signal Processing and Control for Smart Structures

.1 INTRODUCTION

ignal processing and control are both vast subjects. They represent most of the pplied mathematics contribution to electrical and mechanical engineering and now eature strongly in any undergraduate course in most engineering disciplines. It is hen with some temerity that I venture on condensing libraries of evolution into a hapter that may encompass the essentials in a book of this nature.

The aim of this chapter is certainly not to present an up-to-date insight into he nuance and subtlety of modern control and signal-processing techniques. Rather t must be to establish the role and importance of such techniques in smart structures nd attempt to highlight the very basic principles that are involved and that are ommon to most realizations of control and signal-processing systems. Thus, we ffectively define the tools that are available and relate these tools to the problem in and.

The stages involved in the signal processing and control system definition for smart structure are illustrated in Figure 5.1. A very crude division between signal rocessing and control can be defined as a signal-processing function that reduces he data from the sensor array to useful information while the control function, given s knowledge of the structure and the necessary "idealized" structural response, sends rive signals to the actuator array to ensure that this response is closely approximated.

The signal-processing function reduces to two principal phases:

- Representing the input data in an appropriate format;
- Processing this reconfigured data to arrive at appropriate decisions that in turn influence the display (for the sensing structure) or the control algorithm (for the fully responsive structure).

The control algorithm then takes this decision array from the signal-processing ystem and reconfigures it as drive instructions for the actuator array. Of course, the

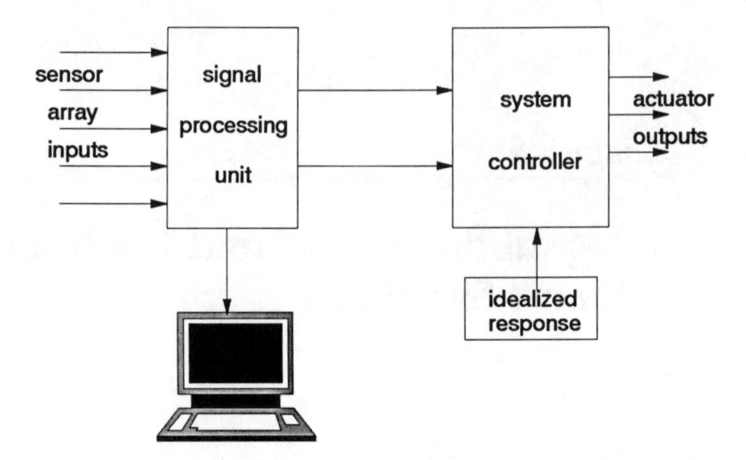

Figure 5.1 Basic functions for signal-processing and control elements in a smart structure.

boundary between the two is extremely blurred, and for the fully responsive structure, the processor and the control elements interact very closely.

The smart structures community presents a very significant challenge in its needs in control and signal processing largely through the shear scale of the problem. The smart structures community lays great store in the biological precursor, and indeed the objective may be encapsulated in the desire for the manmade to be as responsive as the biological. The signal processing community has already taken this on-board, using knowledge-based systems, artificial neural networks, evolutionary computation, natural language control, and a host of other concepts and mapping these into the computer domain. Current limits are imposed by the radical differences in scale between the biological system and its giga or tera neurons and that achievable even on the most densely packed microprocessor chip. At the control end the biological model remains supreme, and manmade output devices remain astonishingly naive. Distributed actuation with efficient energy control and distribution is still something of a long-term dream. Paradoxically this problem makes the problem in the short term rather simpler since dealing with arrays of input information is well within the capabilities of the signal processing community, so at least we can make a start. Dealing with distributed control outputs and multichannel response optimization has, to date, been barely considered.

The scale of the signal-processing and control system in a responsive structure can be immense, so considerable ingenuity is required in the interpretation of sensor array data and the application of the appropriate response. The principal difficulties stem from the combination of the following facts:

- The fact that all sensed data is spatially sampled: Even a distributed sensor can only provide a one-dimensional signal in a three-dimensional structure.

- The structure can require corrective action *anywhere*, and this location should be defined as well as possible, interpolating within the spatial samples.
- But any actuation system apart from a repair instruction can only operate at a limited number of points.

The sensing system should then be capable of interpolating data within the sensor array. Preferably the sensing system should also be passive; that is, it should respond to natural stimuli. In practice the only really successfully fault location systems reported to date have responded to a known, usually mechanical, input stimulus. This is typically a controlled impulse excitation of the structure. The natural stimuli are essentially noise, are almost entirely random, and are certainly—in signal-processing jargon—nonstationery unless viewed over extremely long periods of time.

Ideally too the signal-processing algorithms should have a zoom capacity whereby the occurrence of something unusual should trigger a more refined, high-resolution search of a predetermined volume within the structure. This facility is desirable since a step-by-step analysis of each volume element within a structure could be excessively time consuming. For example, to define the location of an area of interest to within 0.1% of each dimension of a structure would require 10^9 volume elements. In contrast the ideal monitoring system would examine the whole structure at once with a yes:no answer. For the latter the location of the point of interest could be executed on a binary subdivision basis in about 20 interrogations. This ideal scenario does, of course, assume that the single interrogation of the entire structure is capable of providing the necessary yes:no information with sufficient resolution when the location of this point of interest occupies 10^{-9} of the volume of the structure. This gives some indication of the scale of the problem and the trade-offs involved between achievable sensor signal:noise ratio, sensor array complexity, the necessary signal-processing algorithms, and the spatial fault location resolution required within the structure. Of course, if the sensor can locate a fault within 10^{-9} of the structural volume, then the actuator response must also be able to correct to within that volume. The repair man is capable of working to this precision. However, the practical issues of energy distribution and actuator system design have precluded to date the evolution of distributed actuator systems in all except biological species, which in essence distribute fuel to burn at a localized position in order to provide appropriate responses. For biological species the resultant actuation comprises either muscular contraction or the use of some form of localized hydraulics. While the biological actuation system must be admired for its adaptability, it is also significantly hampered in load-bearing capacity and energy transfer capabilities.

The remainder of this chapter presents a brief and, of necessity, superficial summary of signal-processing and control principles. There is a vast and ever-expanding bibliography on the subject and useful expertise on the subjects in virtually all engineering organizations. There is, however, one particular processing feature that is unique to the smart structures area and arises from the variety of sensing instrumentation that is available and permits some forms of spatial integration within the sensing element itself. This is discussed in more detail in the next section, and thereafter we shall review some of the conventional tools utilized within the signal processing and control communities.

5.2 SENSORS AS GEOMETRICAL PROCESSORS

The first stage in processing signals from structures is to detect a *change* in the response signature from that structure. The second stage is to relate this change to structural parameters. Changes are most easily detected if the normal state response of a particular structure is very close to zero. The presence of a signal above this nominally zero threshold is then an indicator that our structure requires more careful examination.

We have already examined at length in previous chapters the types of signals that can be obtained. These may be summarized as:

- A mechanical field, which may be a vibration strain distribution, a static strain distribution, or a map of the propagation of an ultrasonic wave;
- A temperature field;
- A chemical parameter field—typically the location of corrosion products or a safety alarm.

A conventional sensor array comprises a number of strategically located point transducers that measure the value of the parameter of interest and its variation with time at specified locations. An adequate sensor density will ensure that the necessary spatial sampling is in place, and any processing then involves electronically combining the outputs from this sensor array and weighting these outputs to ensure the sensitivity to parameter changes is enhanced. The outputs from each element in the array must be transmitted to a central processor and suitably integrated at this processor. The attendant noise problems and measurement errors must also be incorporated into the processor intelligence.

All sensors in effect take an average of the value of the parameter of interest over the sensitive area presented to the parameter field. Our definition of a "point" sensor is that this area is small in dimensions compared to the dimensions over which the parameter field varies significantly (that is, the shortest spatial wavelength of the parameter field). If the sensor is large compared to the spatial wavelength, then the sensor itself will perform an integration function. The most obvious example of such sensors are optical fiber line integrating sensor systems that can be readily configured to perform beam-forming functions. Figure 5.2 illustrates two simple examples.

Four lowest order mode shapes for a long structure are illustrated diagramatically in Figure 5.2(a) together with fiber optic sensors designed to give optimum

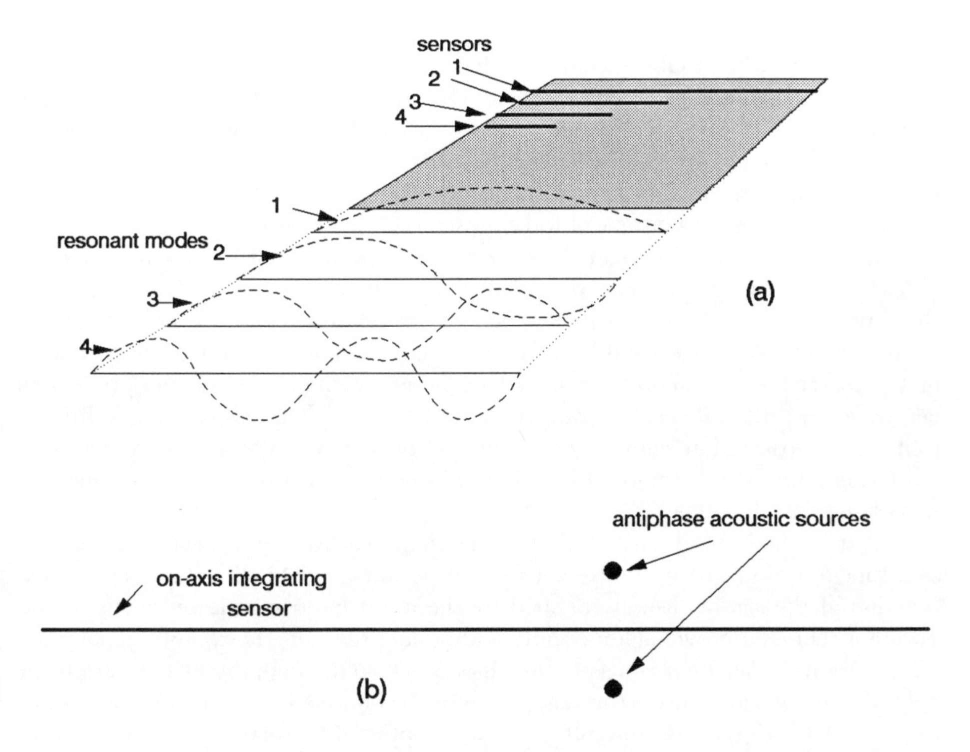

Figure 5.2 Spatial processing using (a) sensor or (b) sensor:source combinations.

response to each of these mode shapes. In this diagram each sensor is designed to be half a wavelength in extent. The position of the sensor along the plate can radically change the relative sensitivity for the various harmonics; for example, if S2 is located centrally it becomes insensitive to even harmonics and only indicates the odd, while if it is as shown in the diagram its sensitivity to the second harmonic is optimized but to higher even harmonics is zero. Applying an impulse to the sample in Figure 5.2(a) to equally excite all modes of vibration will provide different spectral responses from each sensor. These responses are readily calculated and can be made identically zero for a given harmonic by placing the sensor elements symmetrically about one of the spatial nodes. Observing the output at a nulled harmonic from one of these sensors will provide a sensitive indication of the presence of certain types of defect. Anything that perturbs the symmetry of this nulled mode will produce a slight shift in the resonant frequency coupled to a finite output at this slightly shifted frequency from this sensor. This output will be a direct consequence of changes in the mode shape. Therefore, the selective observation of the output spectra from the suitably designed array could be a very useful and sensitive indicator of the mechanical condition of the structure. Such an array has an additional advantage. While, in principle, the ideal situation would involve testing to a known (mechanical) input function, in practice continually observing the ratio of the outputs at a nulled sensor to an optimized sensor (that is, one that is positioned for maximum response to the same harmonic) will ascertain whether the harmonic has been excited (by a random stimulus such as wind, waves, or vehicular traffic). The simultaneous presence of a signal from the nulled sensor will indicate structural changes.

Similar configurations can also be visualized for ultrasonic testing of structures (Figure 5.2b). In this instance an optical fiber integrating detector is placed along the center line between two, equally excited, antiphase fed transducers. There will be an interference null along this line unless there is some deviation in the symmetry of the material—for example, a scattering point—with respect to this predefined center line. Further, if the two transducers are one wavelength apart, an additional null will be observed at right angles to this sensor, again except in the presence of a scattering point or an abnormal load that will perturb acoustic velocity of one side with respect to the other.

A truly distributed sensor—that is, one that measures the value of a parameter as a function of linear position—convolves a parameter field with the interrogation function of the sensor, usually defined by the input interrogating pulse shape and duration. This sensor will again require additional processing to exploit the potential offered by null detection concepts but does allow some mapping of the parameter field. For integrating sensors this mapping must be obtained through inference rather than direct measurement. Suitably configured optical fibers can, of course, perform both the integrating and truly distributed function in one sensor.

The potential offered by integrating sensor systems coupled to suitable signal processing is only now becoming appreciated. The possibilities are illustrated by the calculations in Figure 5.3 that show the relative sensitivities of the four sensor lengths in Figure 5.2(a) to each of the four harmonics (assumed from simplicity to be sinusoidal) as a function of position of the center of each of these sensors. Locations for the various nulls are particularly relevant. Estimating the sensitivity of these null positions to minor changes in the property of the plate is complex in detail. However, estimates of the trends are relatively straightforward and can be obtained from the rate at which the sensor output varies with sensor position around the null and the signal:noise ratio. This rate is at its fastest for the highest harmonics. If we assume a 60-dB signal:noise ratio, then a change in position of the sensor by 10^{-3} of a wavelength will produce a detectable output. This corresponds to altering the symmetry of the structure by 10^{-3} of the highest harmonic wavelength. Whether this is sufficient depends upon the detection criteria to be imposed upon the structure.

The situation for the scatter model is more difficult to ascertain. However, preliminary experimental data based upon probing with Lamb waves of approximately 1-cm wavelengths in a steel plate of the order of 1-mm thick have detected the presence of off-axis scattering points of less than 1/10 of wavelength in dimensions using the line integrating nulling technique outlined here.

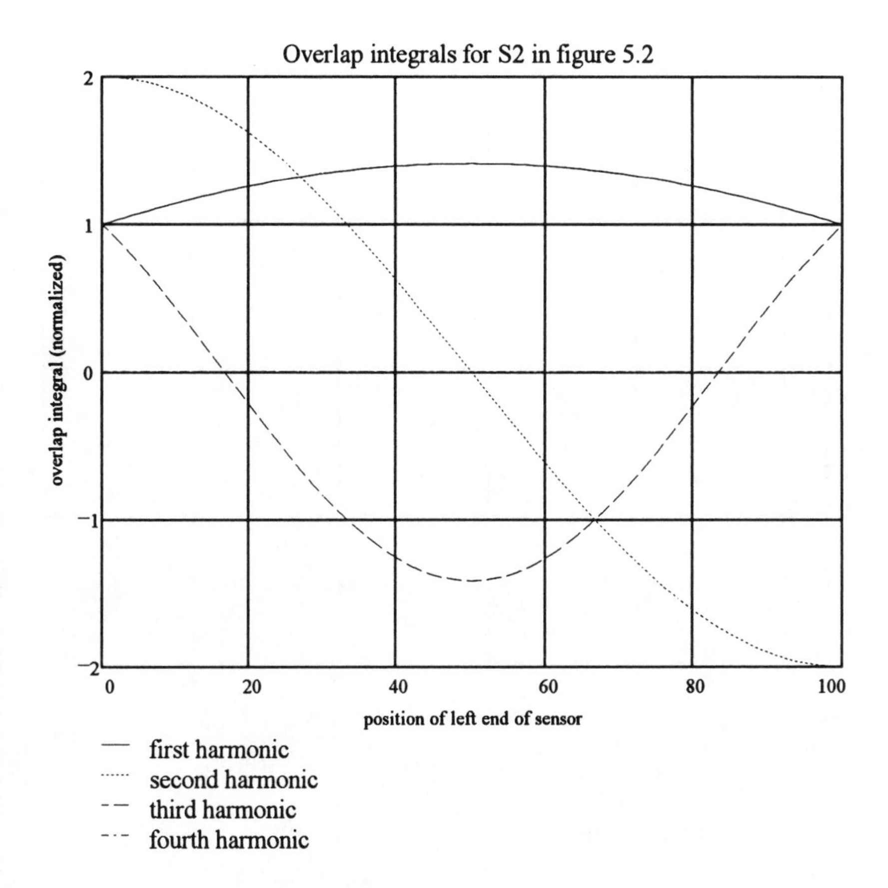

gure 5.3 Overlap integrals for the sensors in Figure 5.2, indicating the spatial integration and filtering capacities of such sensors. S1 in Figure 5.2 overlaps the fundamental spatial mode by a normalized "2", the third harmonic by 2/3, and the zero for the even harmonics. (a) Overlap integral for S2 half-length sensor. Position of sensor is referred to left-hand end: 100 is left end at center of plate. (b) Overlap integrals for S3, one-third total plate width. Position refers to left end of sensor; position =100 refers to left end two-thirds over plate. (c) Overlap integrals for S4 in Figure 5.2 sensor length of one-quarter the plat length. Again the position to left end of sensor; position =100 places the left end of the sensor three-quarters across the plate.

Clearly the use of array processing approaches to sensor design must be specifi-
ally tailored to a particular application. The preceding discussion has indicated that uch preprocessing is not only desirable but may be a very effective approach to ystem design. It also emphasizes the need for a totally integrated approach to esigning a responsive structure. This preprocessing in the sensor array has not yet een formalized into an analytical or algorithmic approach, though the general fea-
ıres of such an approach are straightforward and require integration of a model of the

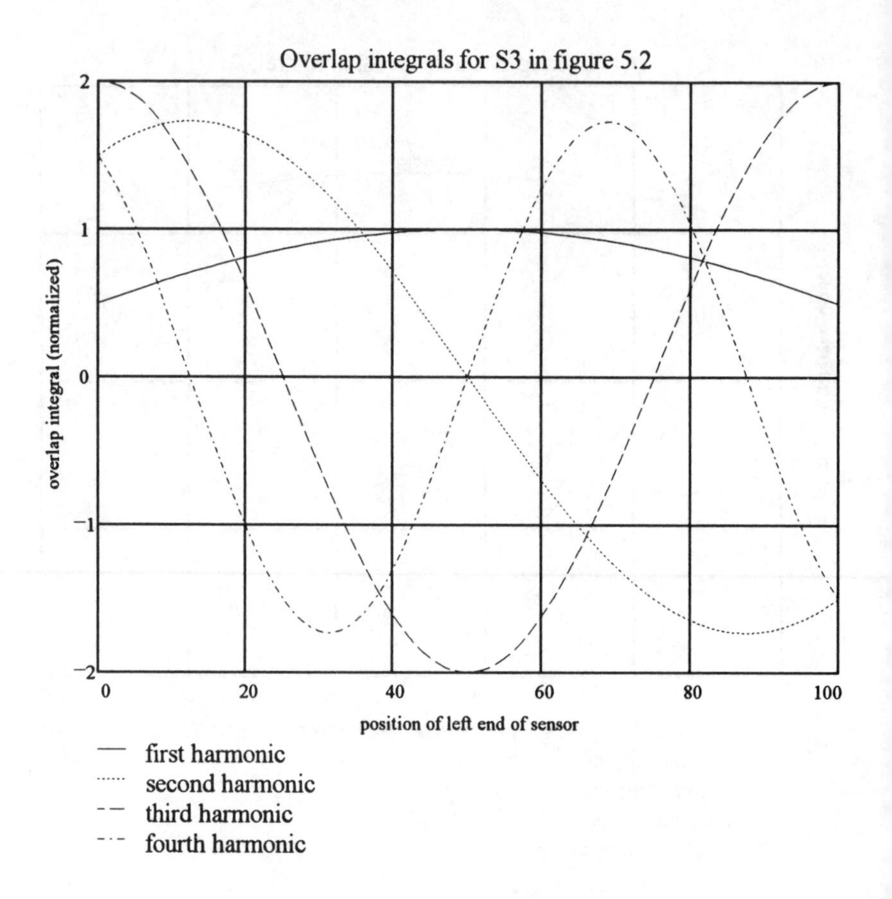

Figure 5.3 (continued).

mechanical response of the system with the response of the sensor to such mechanic[al] stimuli.

5.3 SIGNAL PROCESSING—THE VERY BASICS

We have already recognized the two basic functions of any signal-processing system: data restructuring and representation and decision making. In the vast majority of structural-related applications the input data appear as a spatial representation of the structure and its evolution in time. Probably the majority of decision-making processes based upon this data require a representation of this data as a combination of time and space harmonics, especially when appropriate filtering functions can be applied to either or both domains.

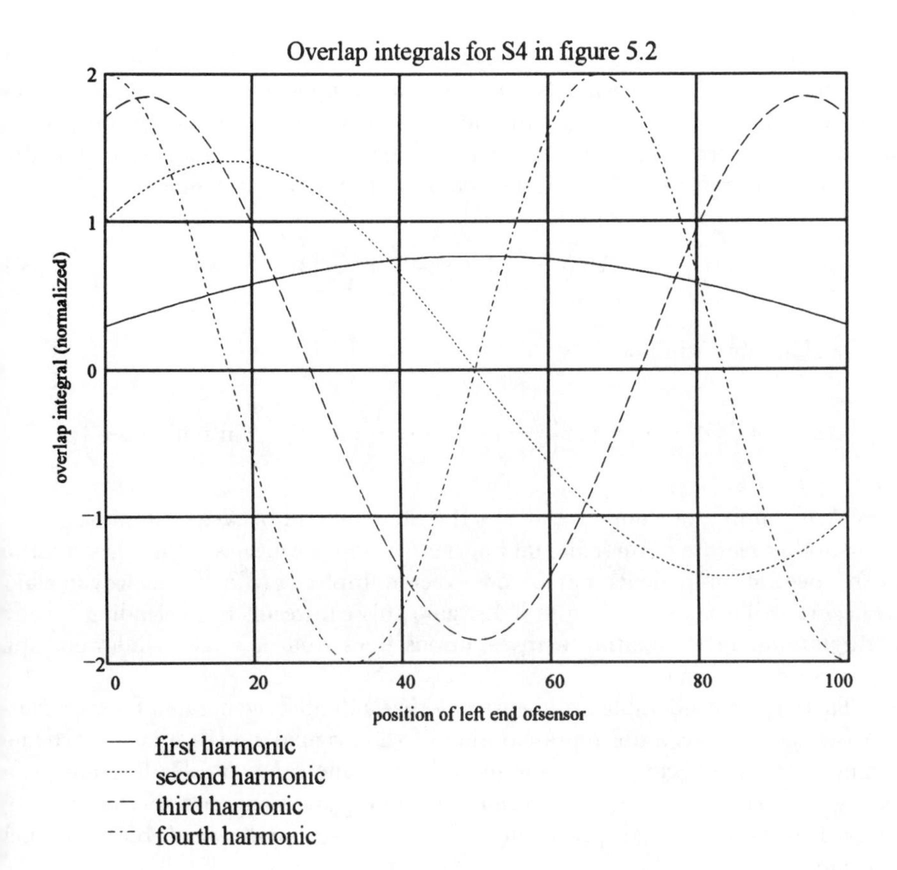

gure 5.3 (continued).

Analyzing signal spectra is a critical operation. The first stage is to represent ıe data as a series of samples spaced by regular sampling intervals in time (though ossibly irregular sampling intervals in space). This operation immediately limits the ıaximum frequency that can be recognized to half the sampling frequency. For the ımporal signals in structural monitoring this rarely if ever presents a problem, but ır the spatial signals it can be a severe limitation especially for parameters subject very rapid spatial variations such as chemical activity.

The sample signal will then be of a finite length—the so-called window—and ıe frequency resolution is always of the order of the inverse of this length.

The windowing operation effectively multiplies the infinite time or space ıquence by a function that is zero outside the window and has finite values within . The impact on the frequency domain is to convolve the spectrum of the infinite ıquence with that of the windowing function. The straightforward rectangular window with its sinc Fourier transform then gives significant spectral spreading an introduces crosstalk between adjacent spectral components. Different windowin functions can be chosen to adjust the side lobes and thereby to modify the spectra spreading—often referred to as "spectral leakage." There is a wide variety of win dowing functions in use, of which the simplest are the raised cosine

$$\varphi(t) = \frac{1}{2}\left\{1 + \cos 2\left(\pi \frac{t}{NT}\right)\right\} \quad (5.1$$

and the Hamming window

$$\varphi(c) = 0.54 + 0.46\cos 2\left(\frac{\pi t}{NT}\right) \quad \left(\text{for } -\frac{NT}{2} \leq t \leq \frac{NT}{2} \text{ in both cases}\right) \quad (5.2$$

where N is number of samples and T is the time interval between samples.

Another feature of these digital Fourier transform routines is that they interpo late for spectral components that are not exact multiples of $(NT)^{-1}$. The way in whic this occurs is illustrated in Figure 5.4. Again this can result in misleading spectra interpretations, often creating many spurious lines from a single windowed inpu frequency component.

There is a considerable art in ensuring that a digitally generated Fourier trans form is a suitably accurate representation of the original signal, and in particula the impact of windowing and sampling are very important: misleading data is s easily generated. The necessary precautions are expanded upon in detail in all th standard texts on signal processing. They are straightforward but extremel important.

The alternative approach to spectral analysis is to invoke so-called "parametri methods" that effectively predict the data outside the acquisition window, producin

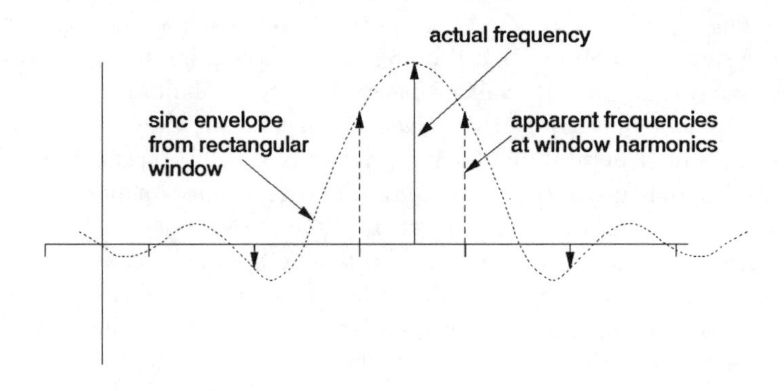

Figure 5.4 Spectral spreading induced by the effect of window functions.

some illusory extension to the sample. The parametric models are to be chosen with great care since again it is all too easy to produce misleading results. All-pole and/or pole-zero models are normally used to model data sets, and as a rule of thumb the order of two or three complex conjugate poles is required to specify each narrowband component in a signal while a wide-band component requires a set of closely spaced pairs of complex conjugate poles. The complexity of the model is determined by trial and error, given some prior knowledge of the spectral components.

There are inherent limitations of the resolution and the accuracy that can be obtained using this process. In particular, closely spaced (compared to $(NT)^{-1}$) frequency components are difficult to resolve.

The concept of stationarity has already been mentioned. A stationary signal is one whose statistical properties remain constant with time. Almost by definition the smart structures processing problem deals with signals that are not stationary. We are, therefore, seeking changes in the statistical properties of the signal and relating these to structural trends. A suitable procedure would then be to divide a continuous stream of data into time cells and examine the structure of adjacent cells. Inevitably these cells must have window functions applied to them, so the sensitivity to data at the beginning and the end of each cell is less than the sensitivity to data in the middle. Consequently, the optimum approach is to use overlapping cells with, for example, the next sample set starting half way through a particular time cell. The data must then be represented as the evolution of a spectral power density with time. Typically frequency will be on the vertical axis while time is on the horizontal axis, and spectral power density is represented through color changes. A profound change in the spectral power density at a particular time is immediately apparent and can be recognized as an event. Again, however, great care is required in defining the window function, defining the overlap function for adjacent windows and matching the window length and sampling into the signal of interest. Exhaustive system simulation, including allowing for known noise sources is essential.

The so-called wavelet transform has recently emerged as a potentially powerful tool to characterize the properties of nonstationary signals. The wavelet transform of signal decomposes this signal into a set of subbands, each of which may be subsampled. The data blocks vary dependently upon the subband of interest, comprising small blocks for low-frequency subband and large blocks for higher frequency subbands.

When the suitable reliable data representation technique has at last been found and proven, the next step is to introduce appropriate modifications to this signal. Highlighting the data of interest is the obvious aim, and it is here digital filtering comes to the fore. Typically digital filtering algorithms act in the time domain and are essentially variations on the delay, weight, and recombine theme to produce a comprehensive array of impulse response functions that in turn can be Fourier transformed to yield a frequency response. Once more there is a vast and ever-expanding literature on this subject, propounding an ever-increasing array of more complex and more elegant routines.

Conventional digital filtering operates with constant delays and constant weights on each delay. A very important variant is one in which these weights and delays may be modified in order to provide an adaptive signal-processing system. In such systems an input signal is compared to a desired signal after processing through some characteristic function. An error signal is derived by tracking the difference between the modified signal and the desired signal-shown schematically in Figure 5.5. The coefficients h_k are optimized from the error signal, which is typically derived from the least mean square or recursive mean square fit of the modified signal x_k, to a desired signal. These adaptive filtering algorithms are extremely powerful and have, for example, been very successful in realizing adaptive noise cancellation and echo compensation in communication systems.

These adaptive filters are simple examples of a linear learning algorithm for which the decision process is based upon a deterministic set of criteria. Within the past twenty-five years or thereabouts a number of very important nonlinear learning systems have emerged. These systems are capable of making judgmental rather than deterministic decisions when presented with appropriate data. They fall broadly into two categories:

- Intelligent knowledge (IKBS)-based systems that compare input data to reference data gathered essentially from human experience;
- Artificial neural networks (ANN) that learn their own sets of rules through training programs that may be controlled by a human operator but are typically based upon internally referenced criteria.

These techniques endeavor to emulate human experience and are typically brought to bear when deterministic techniques prove ineffective. For example, a deterministic technique will predict the modal spectrum of a uniform rectangular plate. This modal spectrum will be modified by damage to the plate, for example,

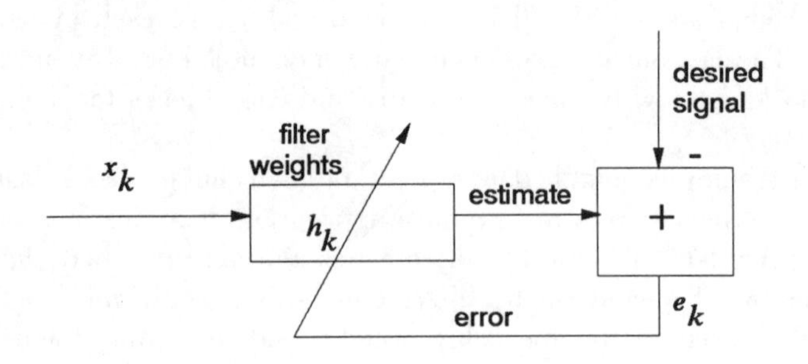

Figure 5.5 Block diagram of adaptive digital filter.

by drilling a small hole therein. However, predicting analytically or through finite element methods the changes to the modal spectrum induced by such damage can be extremely difficult while the experiment is simple. To attempt to locate such damage and estimate whether this damage is waxing or waning is the task for the nonlinear learning algorithm, most typically the artificial neural network.

The general principles of a multilayer perceptron ANN is shown in Figure 5.6. Each input point is connected to all elements in the next layer through weighted interconnects. This layer then triggers through an output function and feeds again through weighted interconnects into subsequent layers. The network may have one or several hidden layers with an array of inputs and an array of outputs. The weights on all the interconnections are adjusted using the learning set to obtain the most reliable interconnection format. This format is never obvious to the user of the ANN. The output is a highly nonlinear function of the weighted input sum and the action of each processing element can be summarized by

$$a = \sum_k o_k w_{jk} \tag{5.3}$$

$$Y = f(a) \tag{5.4}$$

where typically $f(a) = \tanh(a)$ or $f(a) = (1 - e^{-1})^{-1}$ where a is the weighted sum of the inputs from the k neurons in the previous layer weighted by w_{jk} and Y is the

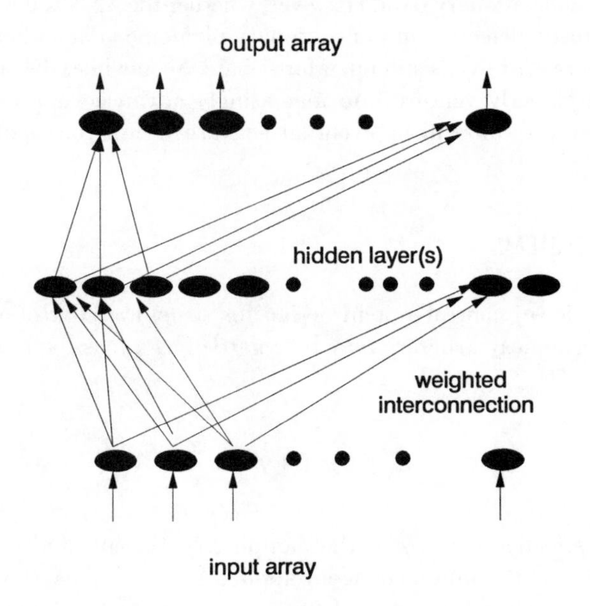

Figure 5.6 General format of a fully interconnect multilayer perception artificial neural network.

output from one particular processing unit as a function of this sum. The hyperbolic tangent and the sigmoid function are effectively thresholding functions that are then passed on to the next layer.

The ANN is based upon the biological precursor. Neurons in biological species behave in a similar fashion. The big difference between the computer and the brain lies in the number of neurons. The human brain packs in around 10^{12} in a highly interconnected matrix that defies analysis. The most complex ANN is very many orders of magnitude less involved.

The fact that the ANN is modeled on the brain and can demonstrably learn operations imbues it with a certain mystique and a consequent fascination. Its ability to learn is constrained by what the ANN is taught, its limited size, and its self-generated conception of right and wrong. It can be a remarkably slow learner, and successful training can take tens or even hundreds of thousands of training cycles. Naive ANNs can be extremely frustrating. There have been many papers published demonstrating that a quite complex ANN can only solve a deterministic problem with 10% of the accuracy of a very simple algorithm in ten times the computation time, so the results do have to be interpreted with considerable caution.

My own belief is that the ANN may be well understood by the computational community but that its ability to register into real life problems is not yet fully characterized. There is no doubt that the ANN will be useful in recognizing structural fault conditions, for example, by comparing the time evolution of signatures effectively by triggering on nonstationary data. However, whether the ANN is the ideal approach when a deterministic detection of nonstationary phenomena is undoubtedly simpler and probably more effective is still open for debate. Meanwhile, the smart structures community will gingerly venture into increasingly advanced approaches to signal processing, and the result will be eventual convergence of the algorithm with the need.

5.4 CONTROL SYSTEMS

The absolute basics of control system design for structural control are very simply expressed. A mechanical structure may be regarded as a mass, spring, and damper combination:

$$F(t) = m\ddot{x} + \mu\ddot{x} + \frac{x}{c} \tag{5.5}$$

where m is the effective mass, μ is the damping coefficient, and c is the effective compliance with $F(t)$ the time-dependent applied force.

We only need a control system if the response from (5.5) is inappropriate, so we modify this to

$$F(t) = (m + k_3)\ddot{x} + (\mu + k_2)\dot{x} + \left(\frac{1}{c} + k_1\right)x \qquad (5.6)$$

where k_3 represents the differential term, k_2 the proportional term, and k_1 the integral term. Thus by applying a force F',

$$F'(t) = k_1 x + k_2 \dot{x} + k_3 \ddot{x} \qquad (5.7)$$

Through a suitable actuation system the structural response can be modified to, in principle, anything that is desired.

Would that it were so simple. There are some important practical problems that are particularly relevant. In structural control:

- The sensor outputs are subject to noise and to distortion and, while in principle all three parameters (displacement, velocity, and acceleration) can be derived from a single measurement, in practice the necessary integration and differentiation routines can introduce significant errors.
- The sensor and the actuator involved are invariably not co-located, so, in fact, the force $F'(t)$ must be applied at one position and its effect assessed with another.
- The control function is often required over a wide area, while current actuators are physically localized; so again some propagation algorithm must be derived.

Of the three terms in (5.7), the proportional term k_2 is, in practice, by far the simplest to apply. This is in effect variable damping, which, of course, implies that in applying this force the actuator must be capable of absorbing or generating energy interacting with the structure. This also implies that the effective resonant frequencies of structures are much harder to control than the damping coefficients. This is hardly surprising since changing effective mass or effective stiffness is intuitively a more difficult operation than absorbing or producing mechanical energy.

Many of the trends in control systems parallel those in signal processing. In particular, the use of adaptive and learning algorithms to optimize system response against established criteria has begun to make significant impacts. Control system design is a well-advanced art and has already made its impact in, for example, active damping of space borne truss structures and sway control in buildings.

5.5 THE LINEAR AND THE NONLINEAR

In mathematical terms the distinction between the linear and the nonlinear systems is very simple. A linear system, when driven by a sinusoidal forcing function at a frequency f, provides an output at that frequency. If the linear system is driven at two independent frequencies f_1, f_2, then the outputs are at these frequencies and no others. Much of Fourier analysis is associated with linear systems, and many of the spectral representation techniques alluded to earlier in this chapter also assume linear systems.

In contrast, the nonlinear system when excited at a frequency f produces outputs at two or more of f, $2f$, $3f$, $4f$, , , . The nonlinear system driven at two frequencies f_1 and f_2 produces outputs at sum and difference frequencies (that is, $f_1 - f_2$ and $f_1 + f_2$) and at sums and differences involving multiples of f_1 and f_2. These lead to the concepts of higher order spectra. The first-order response is a straightforward output:input relationship relating all input frequency drives to the same output frequency responses. Second-order spectra relate the output at sum and difference frequencies and isolate the quadratic term in the response of a system—and so on.

It is tempting to assume that mechanical systems are linear. In practice, it is probably safe to assume that for small excitation about an equilibrium position, mechanical systems in perfect condition are linear. However, mechanical conditions in an imperfect condition are often nonlinear.

The simplest case of a nonlinear system is a beam with a microcrack (Figure 5.7). When the beam is stressed in the direction that opens the crack, its stiffness is determined by the intact portion of the beam. When the forcing function tends to close the crack, the stiffness is effectively determined by the full area of the beam.

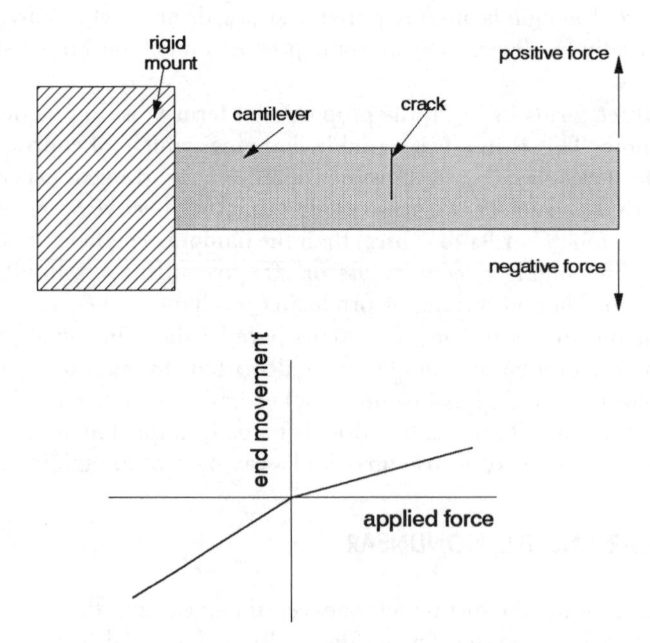

Figure 5.7 Nonlinear behavior of a simple cracked beam.

The stress-strain curve for the beam is then, as shown diagramatically in Figure 5.7, different for compressive and tensile forces. The difference between the slopes is an indicator of a combination of the distance of the crack from the mounting point of the beam and the depth of the crack. Consequently, if we drive the clamped end of the beam with a sinusoidal forcing function (for example, through a vibratory input), the displacement of the end of the beam will follow the form shown in Figure 5.8 with the spectral components indicated.

This very simple example indicates that the onset of nonlinear behavior within a mechanical system—with the appropriate interpretation of the results—can be a very powerful tool in determining the condition of the structure of which the mechanical system forms a part. It does, of course, require a far more comprehensive analysis than that indicated here. However, much of the current research on structural analysis does indicate that the temporal and spatial harmonic contents of modal shapes—of which the above is a very simple example—can be used to an extremely good effect in structural assessment.

5.6 SOME FINAL THOUGHTS

This has been a very rapid glance at two of the most important contributory technologies required in the evolution of smart structures. There are many reasons for this rapid glance, not least of which is that I could hardly claim to be other than an observer in this rapidly evolving domain. Another reason is that the important step is to extract the necessary input:output relationship between a measured structural response and the applied driving action. When this relationship is accurately established, the signal processing and control community is often more than capable of providing an approach to the solution.

In general, there is a hierarchy of necessary functions. The most general is "Is this structure intact?" A yes:no answer to this question determines whether to go farther. The negative response prompts the question "How bad is the damage?" and "Where is the damage?" followed by "What do I need to do to fix it?" While these comments are very obvious it is far from self-evident that they are precisely interpreted by the practitioners. In this regard there is a tendency to overspecify and interpret the necessary data to answer the first question as a full image of the structure's condition. This level of detail is unlikely to be required to answer the last question.

So the temptation to make the overspecified processor must be resisted. This temptation is farther compounded by continually evolving computer technology that offers orders of magnitude more for orders of magnitude less. However, to quote from the Chief Executive of one of the PC suppliers "We give the customer what he wants, not what he needs."

The very positive aspect of the great industry applied to data handling and decision making is that the results are readily available in standard packages for a

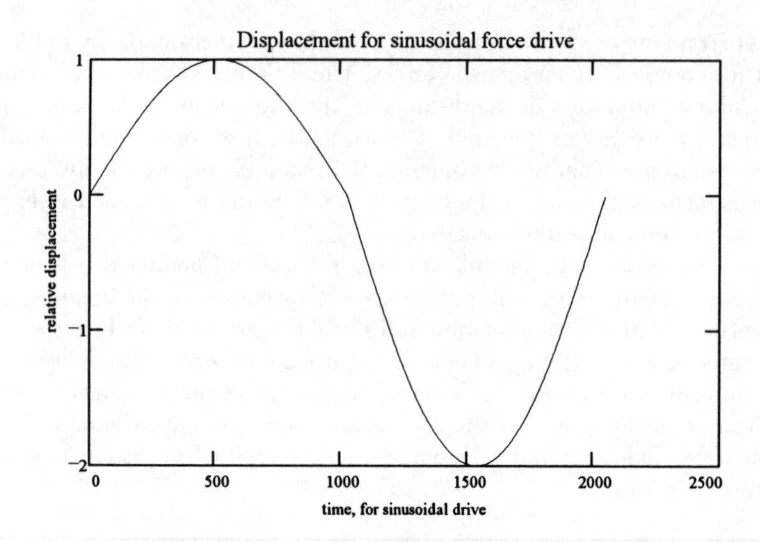

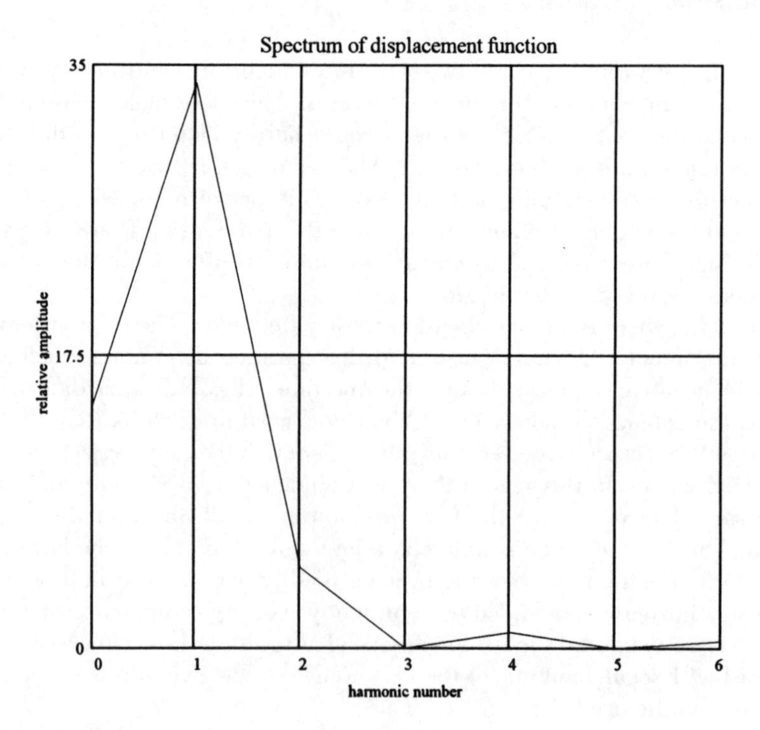

Figure 5.8 The influence of the crack illustrated in Figure 5.7 on the harmonic content of the vibration response to a sinusoidal force drive.

relatively modest cost and that these standard packages will enable the structural designer and user to perform far more functional operations than he will ever find necessary. He does, however, need to understand the very basic features inherent in handling sampled data and the responses that he desires from his network of sensors. He also needs to understand that the sensors cannot produce data that is not there—all that signal processing can do is extract data that is there. Within the preceding few pages I hope to have introduced the basic ideas that are necessary to underpin the signal-processing and control operations.

Chapter 6

Smart Structures—Some Applications

6.1 INTRODUCTION

This chapter introduces some very brief case studies of system usage, exploiting the concepts outlined in the previous chapters. These case studies, which all feature sensing systems and most feature locally controlled reaction, have been chosen to highlight the possibilities that these ideas offer and to indicate some of the current and continuing problems.

The case studies comprise the following:

- Civil structural monitoring systems, including their role in condition monitoring and as potential inputs to the design process;
- Mechanically adaptive buildings with particular emphasis on the control of building sway;
- Active control in space structures for vibration isolation;
- "Smart" composites:mechanical analysis of composite materials and self-testing structures;
- Civil passenger aircraft.

There are several reasons why a "smart" structure could be preferable to an ordinary dumb one. While some of these reasons are more dominant for specific examples within our case studies, there is considerable common ground. In the final reckoning the smart structure must offer some tangible cost:performance benefit when compared to the dumb one. These potential benefits include the following.

- The structural overdesign required to meet extreme loading may be prohibitive and cumbersome, incurring not only substantial cost increases but also aesthetic and utility penalties. This is particularly true for active buildings designed to cope with high winds or earthquake conditions.

- Optimizing maintenance schedules by incorporating intelligence into structures can offer significant benefits. In aerospace this may imply replacing regular maintenance with as needed maintenance and thereby extending the intervals between which the structure requires attention. In civil engineering the basic philosophy is diametrically opposed to aerospace. Often maintenance is only initiated to redress serious situations. Here an early warning system could save billions of dollars per annum.
- The adaptive response of the smart structure will, in principle, enable it to cope with unforeseen circumstances. The dumb structure is only designed to meet the predicted.
- External constraints (particularly weight and cost) impose compromises on the design of the dumb structure so that it is unable to endure absolutely everything that is thrown at it. The smart structure can enhance the range of survivability conditions while operating within the same cost and weight budgets.

The real objective of any smart structure technology is to optimize a life-cycle cost of the structure. In practice, this apparently simple criterion is not so simple to apply, often because the organization responsible for maintaining a decaying structure is not the same as the organization responsible for building and specifying it and, sometimes, even for purchasing it. This is particularly true for situations that incur large but intangible social costs introduced through significant community inconvenience. Virtually all smart, or potentially so, structures involve these intangibles whether for aircraft delays, traffic jams during road repair, or the irritation of the persistent fire alarm. All smart structures—or indeed, all potential applications of smart structures or materials—seem to offer considerable community benefits in addition to those that are quantifiable in terms of direct costs and technical advantages.

6.2 CIVIL STRUCTURAL MONITORING

Civil engineers have been collecting data from structures for a very considerable period of time. In recent years the motivation to collect data on structural performance and structural condition has substantially increased, thanks to a continuing concern about structural reliability and integrity and to a gradual change in design philosophy that has shifted the emphasis from building to a code in practice to building to a specification. In parallel, there have been very significant improvements in data capture technology and computer systems to assist in the interpretation of this data. A commensurate increase in the level of complexity and flexibility in computer-aided design technologies has also contributed to this evolutionary process.

The overall aim of any integrity-monitoring process must be to optimize the lifetime and utility of a particular structure. For example, corrosion mechanisms in bridges are the cause of a few spectacular failures and a vastly greater number of long and tedious refurbishment programs. The cost of such refurbishment procedures is frequently comparable to the initial construction cost. If the inconvenience factor—the local impact of traffic disruption—is incorporated, then the total net social costs are multiplied by a very large factor. Instrumentation that can nip corrosion problems in the bud is potentially very attractive. If the same instrumentation can also influence future design practice to reduce the incidence of corrosion thereafter, then the benefits are even greater.

These potential benefits are obvious but are also purely technical. There are profound cultural factors that also influence the use of instrumentation systems on large civil structures. The civil engineering profession focuses upon large projects that *must* succeed—failures make spectacular headlines. In contrast, the instrumentation industry comprises a diversity of, frequently small, companies dealing in highly specialized areas and is accustomed to solving the specific problems that characterize that particular area. There is an inevitable, sometimes long, learning curve involved in solving these problems during which the instrumentation system almost invariably gives at best inaccurate, and sometimes misleading, answers. The basic philosophies of the two industries are therefore diametrically opposed, so far less has been achieved than could have been possible with compatible philosophies. There are, however, legislative requirements for instrumentation in dams and a few other critical structures in some countries. These can be addressed using well-understood instrumentation techniques.

A monitoring system for a major structure such as a highway bridge poses two fundamental problems to the system designer and installer: (1) How should the instrumentation be interfaced to the structure? and (2) How does one interpret the information that the instrumentation produces?

The latter has a very profound influence on the former. The system may, for example, only be required to comply with safety regulations. Maybe a performance-monitoring system is installed to impose contractual conditions by the customer on the construction industry, or maybe it is added during the structure's life to provide detailed information concerning the progress of suspected fault conditions. The interpretation of information from such systems then is: continue to use the structure, undertake detailed investigations of the condition of the structure at a particular point, and undertake specific repair at a particular point or close the structure.

These global requirements must then be interpreted to answer the first point, namely, how to instrument the structure. The broad parameters of interest may be summarized as follows:

- Measurement of structural behavior, which entails measuring the response of the structure to specific loading conditions. These loads can be dynamic (time constants in minutes or less) or static/quasi-static with time constants in hours or more. In most buildings only static responses are of interest (but see Section

6.3), while in structures such as oil rigs, bridges, masts, and chimneys dynamic responses are also important.
- Measurement of loads that must be known in order to interpret the structural behavior responses. Loads can be directly measured using load cells or by measuring structural strain and assuming a value for Young's modulus. Structural loads are also imposed by winds, waves, and tides; and finally, temperature measurements are critical since thermal loading can impose substantial strains on a structure.
- Measurement of structural degradation, which arises from a variety of causes, the most common of which is deterioration of reinforcement members in concretes, but other effects such as natural erosion, catastrophic movement in foundation position, and material fatigue caused by cyclic loading are also important. The effects of structural degradation can be very serious. While loading can be built into the structural design, degradation is often underestimated.

Structural behavior, imposed loads, and structural degradation can, in principle, all be monitored. In practice, full characterization of all these parameters implies the use of a staggeringly extensive network of sensing elements coupled to careful sensor array data interpretation linked into complex models of structural behavior in order to arrive at useful predictive decisions. Again, in principle, most of the sensing elements required to perform these functions already exist. However, the necessary complex arrays for total characterization are usually unmanageable. The real art of structural monitoring for the civil engineering sector lies in data interpretation and analysis from sparse arrays and in arriving at useful predictive models that permit early corrective action. At present, the industry remains somewhat hesitant in its attitude to such systems because the cost:benefit equations for civil structural monitoring remain difficult to define, especially since the cost of repair is often dominated by the inconvenience factor that is borne by the community at large. Complex instrumentation systems often appear as an unnecessary cost to those who should *directly* pay for them.

The sensing system itself may be designed to perform one or many of the following functions:

- Strain measurement, either throughout the structure as an integrated distance change measurement between strategic points within the structure or as point strains at strategically located points within the structure;
- Crack detection systems;
- Positional measurement of the structure with respect to surrounding fixed landmarks;

- Fatigue cycling monitoring, involving measuring temperature fields and external dynamic loads imposed, for example, by waves and wind (this is particularly important for offshore structures);
- Temperature monitoring, particularly as a correction to strain-monitoring systems;
- Chemical activity measurements and corrosion monitoring.

It is clearly important to select the appropriate measurands from this menu. It is as important to determine exactly where any sensing system should be positioned. For strain measurement the optimum position is clearly in the region of the highest strain. Many structures are based on the cantilever principle, where the highest strain is at the root of the cantilever. However, displacement-sensitive systems are probably best placed either where the movements are expected to be large (that is, at the end of the cantilever) or in positions where unexpected displacements are critical to structural assessment. A vibration measurement using accelerometers should again look at regions of high displacement, whereas vibration measurement using strain gauges should be positioned differently. Temperature gauges are needed to either correct the measurements from strain or displacements or to assess thermal fatigue cycles. Corrosion and erosion are possibly the most difficult measurement problems. Erosion almost inevitably occurs on the outside of the structure and so can be, at least in principle, monitored through visual inspection. Corrosion can occur anywhere and, since it often affects strength members, can be critical anywhere. To date corrosion has often been inferred as an interpretation of anomalous displacement measurements. An anomalous displacement almost always implies that the underlying reason is serious and will require extensive corrective action.

The requirements within monitoring systems for civil structures is encapsulated in Figure 6.1. There is clearly a need for a very close interaction between the design of the sensor system, the civil engineering features of the structure, and the decision processes that may be fed to either or both of the owner of the structure and the regulatory authorities.

The Kingston Bridge in Glasgow is often claimed to be United Kingdom's busiest highway crossing, carrying over 150,000 vehicles each day. It comprises two separate spans, each carrying five traffic lanes (Figure 6.2), and was built about twenty-five years ago. The main span is 145m with side spans each about 63m, leading to approach roads totaling some 1.5 km in length. In addition to being United Kingdom's busiest bridge, it is also United Kingdom's most measured. It suffers from what in the popular press is referred to as "concrete cancer," arising from severe deterioration of the reinforcing members within the structure, which features post-tensioned concrete-in-a-box section format.

The instrumentation on this bridge [1] (Figure 6.3) comprises the following:

- 36 LVDTs with 100-μm resolution located in the expansion half-joints on the top and bottom pier bearings and at the north and south ends of the main side spans at the expansion joints to the approach roads;

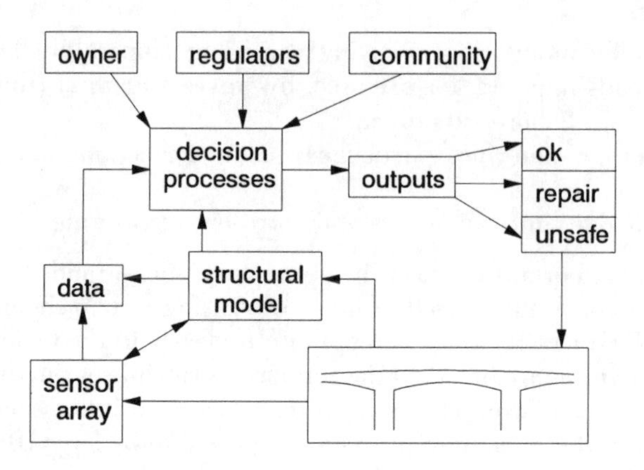

Figure 6.1 Monitoring civil engineering structures—basic elements.

- 128 platinum resistance thermometers located at various depths and locations in the concrete of the structure and within the internal voids;
- 16 corner cube reflectors, 8 each on the east and west outer faces of the bridge plus an additional 8 on the soffit of each bridge structure at the north pier, aligned to laser range-finding survey stations located on the foundations of the north and south piers;
- Weather stations above and below the bridge monitor temperature, relative humidity, wind velocity, and direction during measurements.

The stimulus for this extensive instrumentation package was slippage in the north pier foundations observed in the late 1980s. The quayside beneath the north pier bulged by up to 200 mm, and the north pier itself deviated from the vertical as a consequence. Further instrumentation was, therefore, installed within both north and south quayside side walls. This comprised six pairs of retroreflecting prisms on the north wall above high-water mark and three pairs on the south wall. In addition, tilt meters were installed in the north quayside and inclinometers on both north and south quay walls. This instrumentation was primarily used to monitor the necessary repair works, including the application of some 16,000 tons of boulders against the north quayside wall and a jet grouting process used to directly strengthen the foundations. Figure 6.3 shows the positioning of the major instrumentation features in the bridge structure. The bridge instrumentation is interfaced to automatic data-logging equipment, and the LVDT suite and the array of resistance thermometers can be continuously monitored. The retroreflectors are surveyed using laser range-finding equipment as necessary. Some typical results are shown in Figure 6.4, including (a) the changes in positions of prisms on the north wall during the refurbishment of the foundations and (b) a typical plot of an LVDT versus time plotted with air

Figure 6.2 Photograph of the Kingston Bridge, Glasgow. (*Source:* Kajima Corporation. Reprinted with permission.)

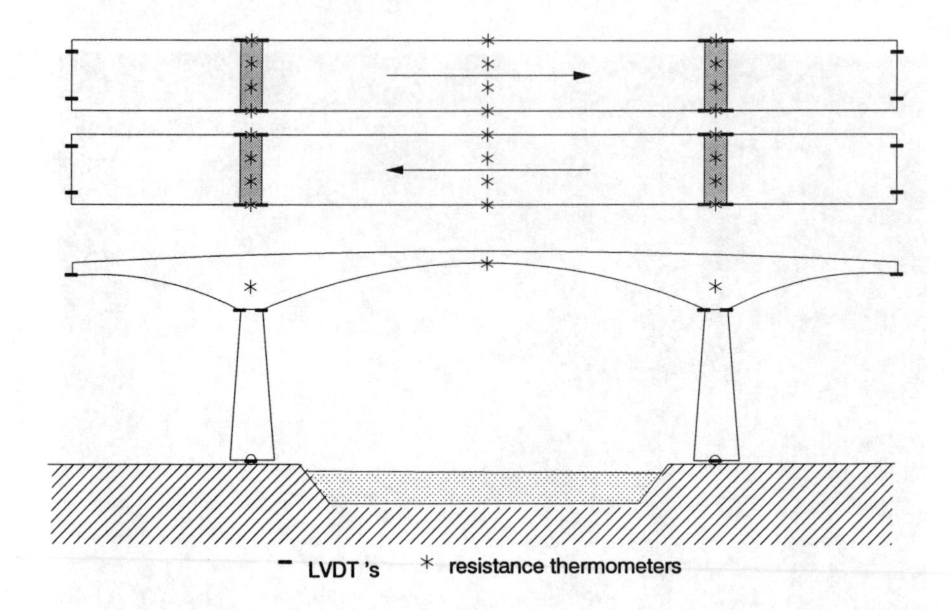

Figure 6.3 Indicative positions of sensors on the Kingston Bridge.

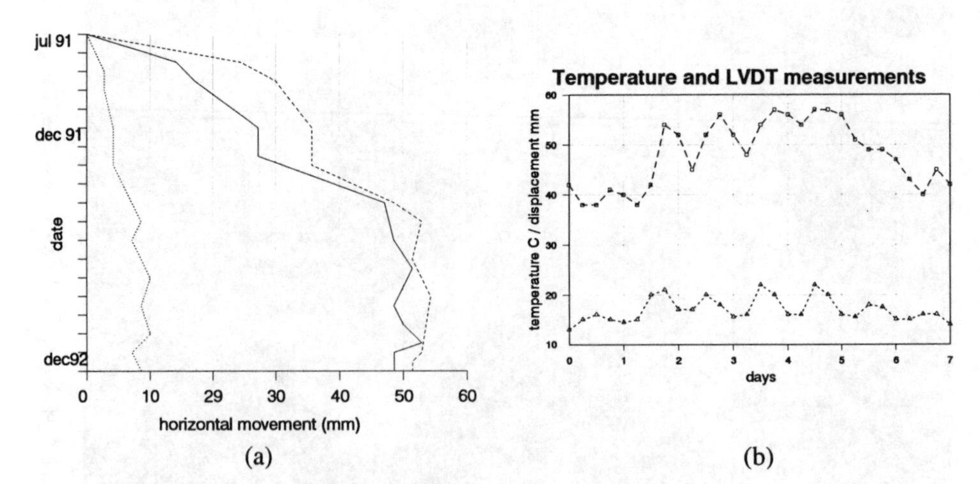

Figure 6.4 Typical measurements from the Kingston Bridge illustrating (a) structural motion during refurbishment (see text) and (b) climatically induced changes.

temperatures in both the surrounding air and the void within-the-box section span of the bridge.

These monitoring systems continue to be heavily used after half a decade. Further extensive repairs are imminent, and the system will also provide indispensable information concerning structural movements during future refurbishment. The installation of this measurement system responded to a real problem; however, the system has also demonstrated its potential to indeed furnish the necessary early warnings.

6.3 AN ACTIVE BUILDING

The term "intelligent building" is much abused and usually implies extreme (in my opinion unhealthy) control of light, space, and air. The active building is an entirely different concept. Here it is the mechanical properties of the building that the designer seeks to modify in response to external stimuli. The idea is particularly apt in regions where the probability of earthquake is high but is also very useful in areas where there are frequent high winds. The overall idea is to minimize the sway of a building in response to external forces such as ground movements and air currents in order to maximize the comfort and safety of the occupants.

Active buildings have been pursued most energetically in Japan, where several tens of buildings have already been equipped with specifically designed vibration-damping systems. The objective of all these systems has been to minimize the response of the building to wind or earthquake stimuli.

The active building systems are based upon one or a combination of the following principles:

- Active mass drivers in which a large mass, typically several tons in weight, is mounted on rails on the roof and driven along these rails to counteract the effects of any building motion; typically two orthogonal path mass systems are required;
- Tuned mass damper systems, which are passive and involve the design of a heavily damped mass:spring oscillator with a resonant frequency equal to the building resonance and coupled to extract vibrational energy from the building, again usually mounted on the roof;
- Active variable-stiffness systems, which serve to decouple building resonances from ground vibration frequencies to minimize coupling effects, thereby "reflecting" mechanical energy from earth tremors.

Table 6.1 [2] summarizes one set of such active buildings. The active mass-damping (AMD) systems and hybrid mass-damping (HMD) systems comprising both active and passive components are designed to minimize vibrational amplitudes in

Table 6.1
Some Adaptive Buildings in Japan

Name of Building	*Completion (Scheduled)*	*Building Description*				*Type of Control System*	*Type of Disturbance*
		No. of Floors	*Max. Height*	*Total Floor Area*	*Type Floor Area*		
Kyobashi Seiwa Bldg.	Aug. 1989	11 + 1(BG)	33.10 m	423.37 m^2	37.32 m^2	AMD system	ME, SW
KaTRI No.21 Bldg.	Nov. 1990	3 + 2(BG)	16.30 m	465.00 m^2	150.00 m^2	AVS system	LE
Ando Nischikicho Bldg.	July 1994	14 + 2(BG)	68.00 m	4.928.30 m^2	324.15 m^2	HMD system	ME, SW
Shinjuku Park Tower	Apr. 1994	52 + 5(BG)	232.60 m	264,140.91 m^2	4,523.54 m^2	HMD system	ME, SW
Dowa Kasai Pheonix Tower	Jan. 1995	29 + 3(BG)	144.45 m	30,369.66 m^2	1,072.36 m^2	HMD system	ME, SW

BG, below ground level; ME, moderate earthquake; LE, large earthquake; SW, strong wind.

wind speeds of up to 20 m/s to one third of the uncontrolled response and to limit vibration amplitudes thereafter.

The KaTRI No. 21 building in Tokyo is shown in Figure 6.5. This building has large active inverted V-shaped braces installed in each story. These braces may be switched from high-to-low stiffness states by opening or closing a valve placed inside the device. The building is designed to withstand large earthquakes. The general principle is to measure the earthquake acceleration characteristics, analyze the motion through bandpass filters, and from this analysis extract the optimum building stiffness from the options available by switching in or out the various variable-stiffness devices. The decisions to open or close the valves in the twelve stiffness devices are then made and the structural stiffness change implemented. The whole process takes less than 40 ms. The coupling of the building to the applied mechanical movements can then be very rapidly varied. However, the ability of the structure to dissipate the mechanical energy that is coupled into it is essentially unchanged. In contrast, active and hybrid mass-damping systems are designed to extract mechanical energy from a swaying building and dissipate as much of this energy as possible in oil dampers or similar arrangements as heat. In passive mass-damping systems the tuned mass is typically 1% or less of the total building weight, so the effective stiffness of the mass-mounting system is in a similar ratio to that of the overall building. The tuned mass acts as a damped coupled oscillator, and typically two systems are required to cope with oscillations along the two principal axes of the structure.

Active mass systems effectively do the opposite of the child on the swing (that is, pump inward at the end of the swing) to extract mechanical energy from the building into the active mass and hence to dissipate this energy as heat in the drive mechanism.

In hybrid systems the tuned passive mass damper is augmented by an active system that typically involves moving weights of the order of 20% of the total mass of the passive damper. Figure 6.6 shows the Dowa Kasai Phoenix Tower in Osaka that is equipped with two hybrid mass-damping systems located at the top of the building (Figure 6.7). This system reduces the response of strong winds of up to 15 m/s (which are anticipated no more than once every 5 years) to one-half of the uncontrolled response. This in turn reduces to an imperceptible level the seasick-like feelings of nausea and discomfort that would otherwise be experienced by the occupants of the building. The mass damping system (Figure 6.8) comprises a tuned mass of 30 tons and two active mass dampers of 6 tons each.

The building has recently been completed, but during the design stage extensive wind tunnel experiments were conducted, from which the control response can be predicted with considerable confidence. A plot of the acceleration response under strong wind excitation is shown in Figure 6.9, which demonstrates a typical reduction in acceleration amplitude of a factor of 3.

During their relatively short history, these buildings have already provided a wealth of information concerning their response under both wind and earthquake

Figure 6.5 The KaTRI No. 21 building including overall composition of the AVS automatic variable stiffness system [2].

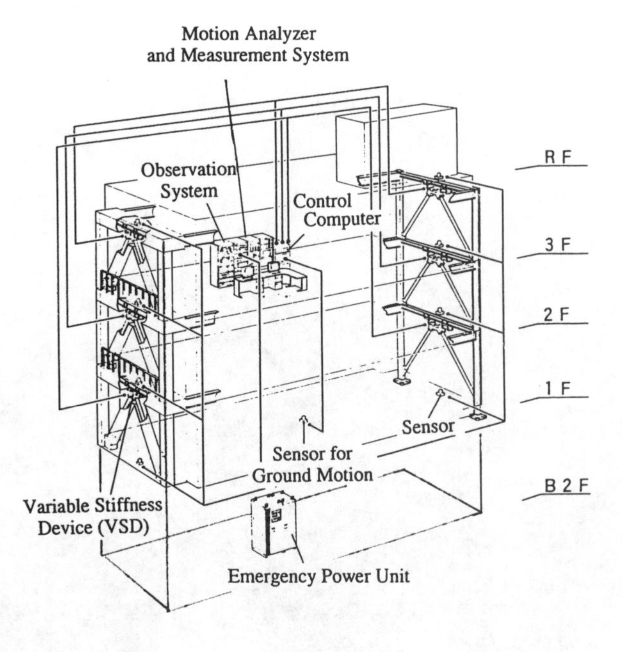

Figure 6.5 (continued).

excitations. In most cases the observed response with the control systems switched on has been compared with a simulated response, but in a few, immediate before and after comparisons have been available. Typically factors of 2 to 3 reduction in acceleration amplitude within the building have been observed with commensurate reductions in mechanical energy coupling of approaching an order of magnitude. In the AMD and HMD systems the energy difference is dissipated as heat within the damper, whereas in the variable-stiffness device systems, the objective is both to discourage mechanical energy coupling and to distribute the dissipation of any coupled energy as harmlessly as possible.

Finally it is interesting to compare the problem of vibration damping in buildings to that of vibration damping in vehicles. In the latter vibrating panels are usually inhibited by applying actuators such as piezoceramics to localized high-strain regions and dissipating the energy in an electrical load connected to the piezoceramic. The equivalent in the building would be a piezoceramic foundation layer or, equivalently, an array of hydraulic jacks. The moving mass systems are in practice very significantly more straightforward in their implementation.

6.4 ADAPTIVE TRUSS STRUCTURES

The space structures community makes extensive use of structures based on networks of interconnecting rods rather like an engineered version of the childhood construction

Figure 6.6 The Dowa Kasai Phoenix Tower, Osaka. (*Source:* [2]. Reprinted with permission.)

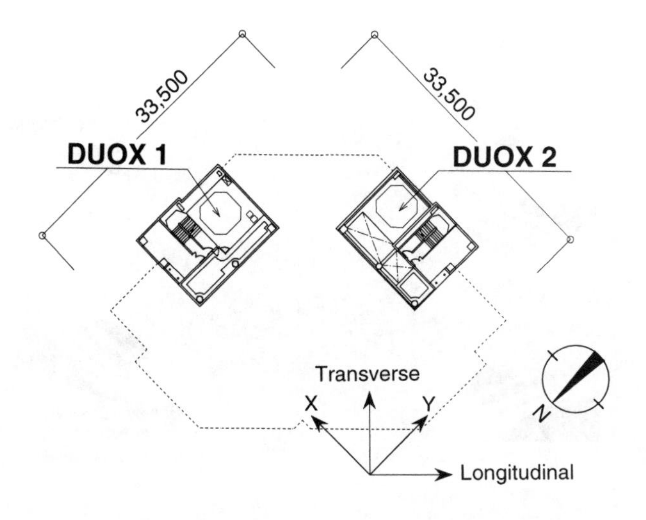

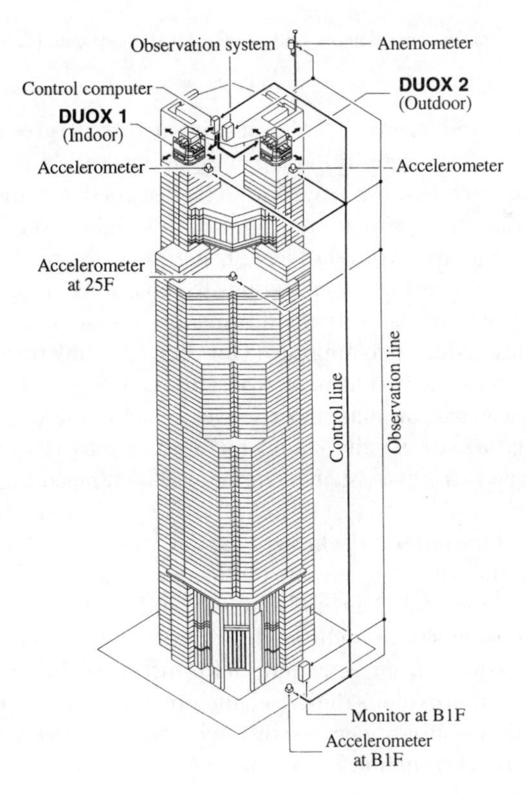

Figure 6.7 Schematic showing location of sensors and DUOX combined active and passive mass-damping systems. (*Source:* [2].)

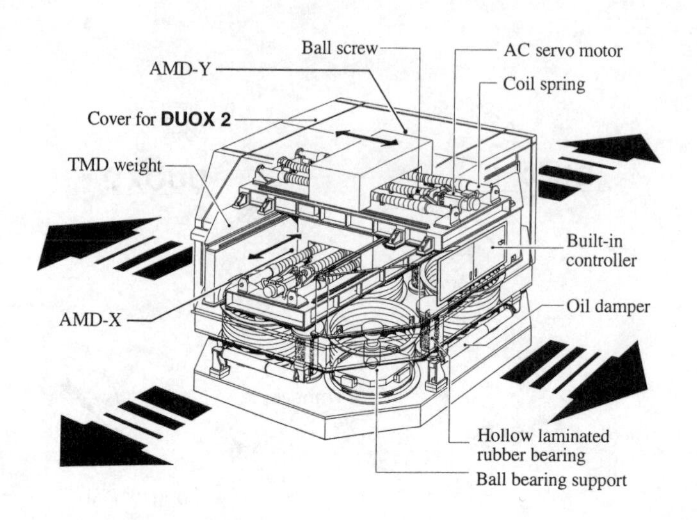

Figure 6.8 Schematic of active mass damper and tuned mass damper unit. (*Source:* [2].)

set based upon drinking straws. The interest in such structures is readily explained. Stiffness (or more strictly, rigidity) is far more important than strength, and weight is a primary consideration. An appropriately designed arrangement of rods and interconnecting triangles optimizes the rigidity to weight ratio (Figure 6.10).

Such rigid structures are relatively straightforward to design and, especially when constructed from carbon fiber composite rods, have an extremely high stiffness:weight ratio. They are also very good transmission media for vibrational stimuli, so active systems are required to minimize vibration coupling through such structures. These are usually implemented using active strut members within the truss, typically based upon piezoelectric or magnetostrictive linear actuators (Figure 6.11). The piezoelectric actuators are ideally suited for applications requiring relatively short stroke; while magnetostrictives, such as terfenol-D, can be configured into somewhat longer stroke systems [3].

An example of the latter is the hexapod active vibration system (HAVI) described in [4] based upon the so-called Stewart platform concept. Each of the six legs of the hexapod features a terfenol-D linear actuator, and all the legs are arranged to be continuous and to intercept at right angles (Figure 6.12). The intent of the system is to isolate one platform from vibration, fields affecting the other.

The control system exploits the impedance-matching concept described in Chapter 4. The vibration source operates through the structure's impedance into the actuator impedance. A function of the actuator is to match its impedance to the complex conjugate of the structure's impedance, thereby optimizing energy transfer

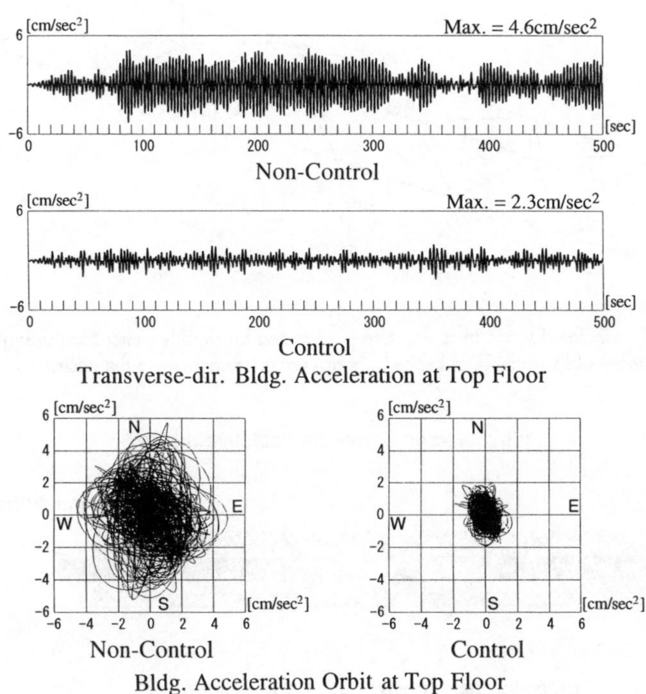

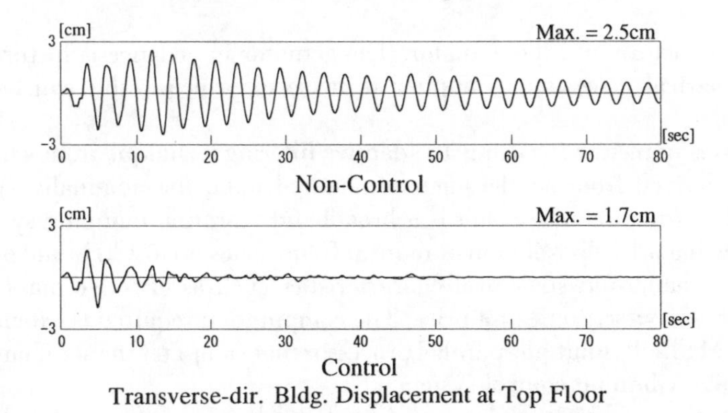

Figure 6.9 The control effect introduced by the DUOX system. (*Source:* [2].)

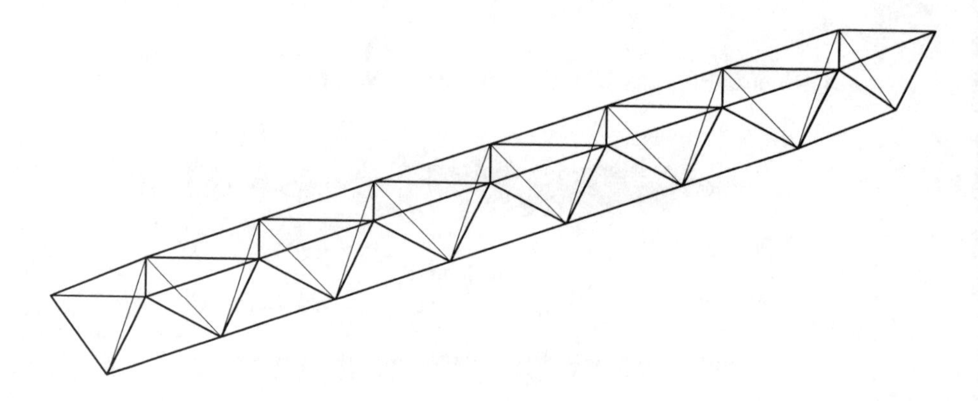

Figure 6.10 Triangulated space truss structure—optimized for rigidity rather than strength. The length tolerances of the individual brace bars are critical to achieving the necessary rigidity.

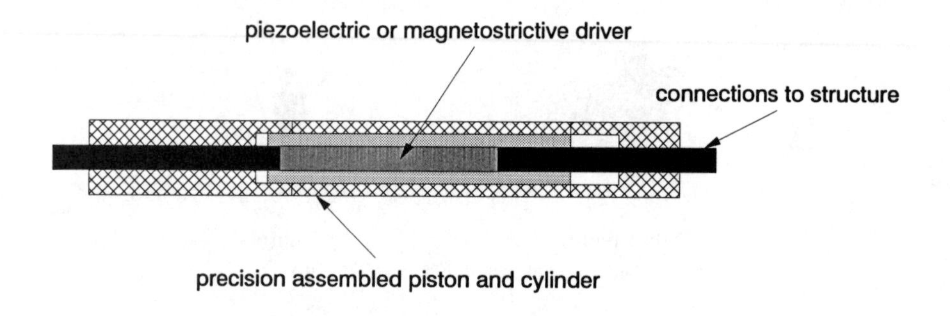

Figure 6.11 Active truss member for space structures—simplified schematic.

from the structure into the actuator. The actuator impedance is in turn a function of the electrical drive that is applied to it, so in principle this can be electrically controlled.

This is implemented using an adaptive filtering technique from which the error signal is derived from accelerometers mounted upon the nominally vibration-free part of the structure. The result is a broadband vibration-damping system capable of introducing a 15-dB isolation or more at frequencies up to 1 kHz and of responding rapidly to changes in structural characteristics such as effective mass or stiffness. This does, of course, come at a price. The computation required is extensive running on a 566-M FLOPs multiple-parallel processor that computes the six-channel simultaneous active vibration control system.

A very considerable volume of knowledge has been accumulated both in the United States and Europe on the properties of such structures. They serve to illustrate a significant achievement that may be realized by integrating advanced actuator

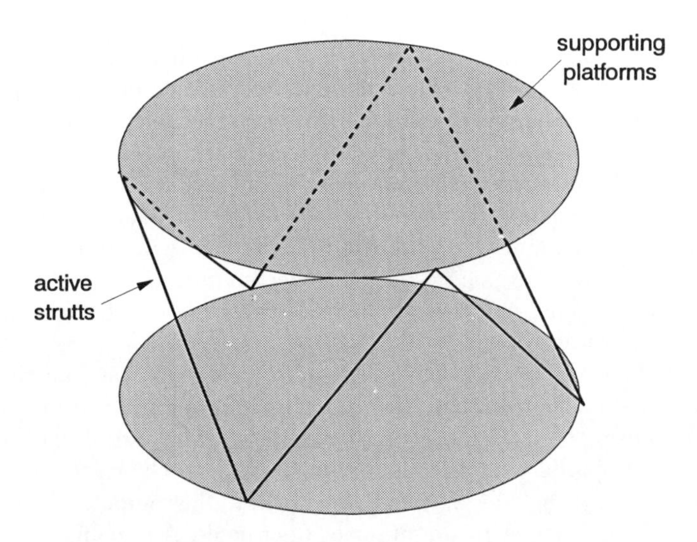

Figure 6.12 The hexapod or Stewart platform—all actuators intersect at right angles.

materials, state-of-the-art structural materials, up-to-date control and signal-processing algorithms, and extensive computational facilities. These structures are not specifically designed to be load bearing, so their direct terrestrial applications may be limited. However, the principles that underlie their design and implementation offer considerable potential for the more commonplace environments for which currently, while the performance is desirable, the cost is prohibitive.

6.5 "SMART" COMPOSITES: MECHANICAL ANALYSIS AND SELF-TESTING STRUCTURES

Carbon and glass fiber-reinforced plastic materials have been known for decades and recognized for their exceptionally high strength:weight ratios. They have been used for body panels, suspension systems, wing sections, tail planes, and storage tanks for many years. They are even beginning to emerge as serious contenders for reinforcing elements in concrete structures and as the structural members in highway bridges.

And yet, with all this experience, there continues a professional unease with the application of these structures especially in safety critical areas. There are probably two principal reasons for this: (1) the material is essentially inhomogenous, so its detailed structure especially in thick sections is often totally unknown; and (2) there is always the fear of voids or resin-rich areas that may significantly affect the overall strength of a particular structural member. This fear is exacerbated by the knowledge that the properties of the resins are usually cure process-dependent, and in particular, in thick structures, this is often an unknown process. The situation is made yet more uncertain by the use of epoxies to join large structural members together—for example wing sections in aircraft—with the normal human suspicion of a glued joint that is never felt to be quite the same as one that involves mechanical connection, despite the fact that the latter may well be less reliable. Second, the in-service fatigue wear and damage characteristics of these materials are still uncertain. The user is particularly concerned with delamination damage—which may be introduced by, for example, dropping the airman's hammer on the wing panel—that will not be readily visible at the point of impact. Fatigue cycling is better characterized, but the effects of solvents and, in particular, moisture on the performance of the all-important epoxies remain somewhat uncertain.

Much of the early work on so-called smart structures was motivated by the possibility that these problems could be solved by appropriate instrumentation and in particular by using optical fibers as sensor elements. The optical fiber was viewed as being totally compatible with the reinforcing fibers, particularly in terms of ability to withstand strain cycling and to be embedded within the composite with impunity. Furthermore, it was felt that this omniscient fiber could also monitor the cure and fabrication processes as well as in-service operational characteristics.

A very great deal of work has been carried out to determine the compatibility of an optical fiber sensor with a composite with the overall conclusion that optical fibers may indeed by embedded within both glass and carbon composite materials and provided the process is done with sufficient care there will be little impact upon strengths, breaking strains, or overall mechanical properties. It has been found to be critical that the optical fiber is oriented appropriately with the reinforcing fibers, that the optical fiber is less than 10% or thereabouts of the thickness of the composite in diameter, and that the optical fiber is coated with a resin-compatible material, typically a polyimide (Figure 6.13) [5].

A multitude of techniques have since emerged for measuring strain and temperature in composite materials using optical fibers either in distributed, quasi-distributed, or point (usually Bragg grating) sensor formats. The result has been, for reasons that have already been discussed in this book, a reaffirmation of the need for simultaneous strain and temperature measurement for quasi-static measurements coupled to a reliable thermomechanical model of the material. The process has also stimulated serious questioning of the value of load:strain relationships and vibration characteristics without prior knowledge of both the detailed characteristics of the load and temperature conditions. Perhaps with hindsight the composite community needed to endure the learning curve that their colleagues in the civil structures community had progressed along already to find that comprehensive data is necessary in order to recognize structural anomalies.

Manufacturers of composite materials continue to highlight the need for processors that will monitor the curing cycle of the epoxy resins during manufacture despite over a quarter of a century of practical and largely successful experience. Engineering attempts to characterize the curing process have been largely inconclusive despite

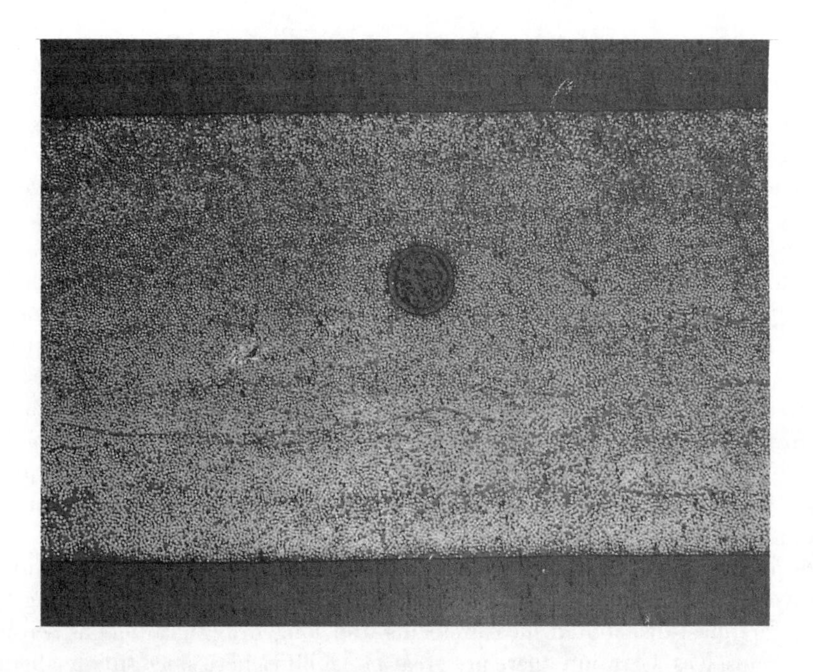

Figure 6.13 Photograph of optical fiber embedded in carbon fiber composite to be used in structural monitoring.

the fact that many physical and chemical phenomena undergo marked changes within the cycle. There are significant changes in dielectric constants, ultrasonic propagation characteristics, Raman spectra, refractive indices, and mechanical properties that can, in principle, be used. All of these techniques—and many others—have been evaluated and have certainly presented recognizable signatures that characterize the continuing curing process. However, none have, as yet, gained broad acceptance, principally because differentiating between a "good" cure and an "unacceptable" cure within the general trends of the available signatures has, as yet, to be reliably demonstrated. The need persists and will succumb to the appropriate combination of sensing techniques and data interpretation. In the sensor domain an appropriate combination of sensitivities to chemical and physical conditions is essential since both are required to establish confidence in the material structure. Data analysis is arguably more difficult since the trends within the sensor signatures must be distinguished in order to differentiate between the good and the unacceptable. To realize this requires samples of both states and confidence that simulations of the unacceptable are realistic representations of practical situations.

The process is ongoing. The fiber sensor research community doggedly continues to improve the basic technology by, for example, using short Bragg gratings as reflectors for quasi-distributed measurements and long Bragg gratings as sensing elements themselves. Certainly there are great possibilities here since simply aligned sensor arrays may be readily realized and such arrays are readily compatible with direct integration within the material.

The complete-systems approach to the interpretation of the results is critical to final success, and it is here that passive measurement systems in composites have made but little progress. There are undoubtedly several parameters that could be measured, but there is no ready algorithm set to interpret these measurements.

The basic problem here is that a sensing system is measuring the structure's response to an undefined stimulus. The obvious progression from this observation is to define a stimulus. The entire science of nondestructive testing works on the basis of interpreting deviations from standard response to a standard input in terms of structural parameters. Indeed there is a very comprehensive literature on conventional NDT in composite materials. In a different regime there is also a comprehensive "tap and listen" testing tradition of which the most obvious is the railway wheel tapper whose basic skills have been applied to similar problems in boilers, steam pipes, ship hulls, and automobiles. Again the basic operation is the interpretation of a response to a known stimulus.

So why not automate the wheel tapper? This operation enables the recognition of the *presence* of a fault rather than the detailed characterization of the fault that typifies NDT. However, for many, possibly most, requirements this first step is the critical one in determining whether or not a particular structure remains serviceable.

There have been many attempts at this process, some in composites and some in more conventional materials. Some have been encouragingly successful. At one extreme there is the impulse excitation of a bridge in New Mexico that proved to be able to locate deliberately introduced damage in the structural members with remarkable precision [5]. In contrast, basic experiments using impulse excitation of simple plates both of composite material and metals have highlighted the very critical dependence of the modal response on structural boundary conditions. For many structures these boundary conditions are impossible to maintain, so modal analysis as a general inspection procedure is probably unacceptable even though it is highly sensitive to structural changes if the boundary conditions do remain constant For some applications, analysis of modal shapes, and in particular detection of modal shape curvatures, especially discontinuities therein, can provide useful information. This analysis can highlight, for example, cracks. However, for graceful decay, for example, through gradual corrosion, the generation of mode shape discontinuities is somewhat less predictable.

The inevitable conclusion is that any probing technique should be preferably independent of structural boundary conditions. Conventional NDT is one such example but is too complex for routine application. However, it does lead quite naturally into the ultrasonic wheel tap shown schematically in Figure 6.14. This highlights some interesting properties for ultrasonic fault detection in composites. Defect sizes of the order of 1 mm should be detected in any fault alarm system. However, the acoustic propagation velocities are such that a 1-mm wavelength, which is required for

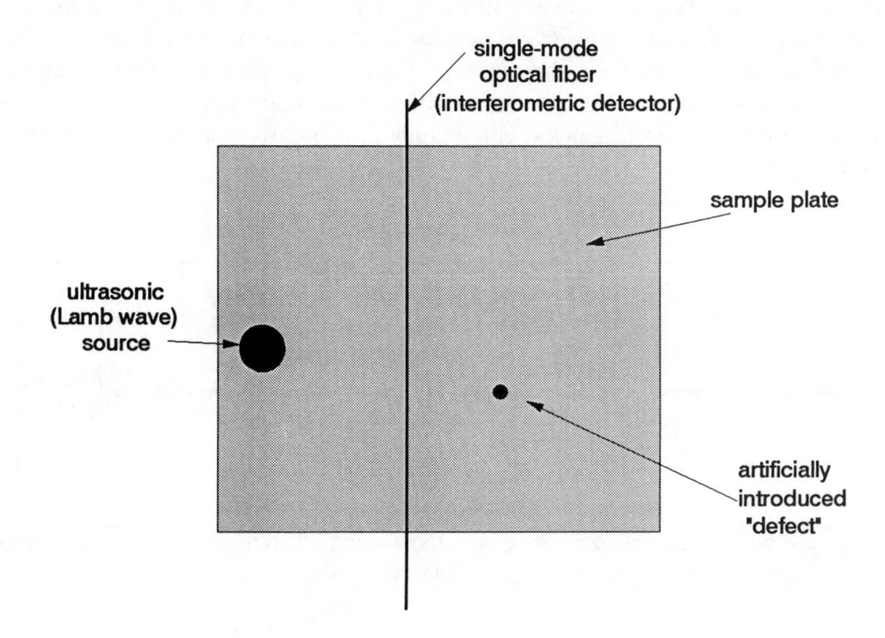

Figure 6.14 Defect detection using optical fibers and long wavelength ultrasound.

conventional imaging of such a defect, corresponds to excitation at several megahertz, where the losses found by most workers are excessively high, that is, in the region of decibels per centimeter, precluding examination of even moderate-sized structures. Wavelengths of ten times this value suffer relatively little attenuation but can only detect the presence of such small defects by scatter rather than imaging reflections.

This concept as a composite evaluation tool is in the very early stages of its development. However, the trace shown in Figure 6.15 shows the signal detected using an optical fiber outside the direct (ultrasonic) illumination zone from the ultrasonic source as a result of a very small scatter of dimensions of the order of 1 mm. This proves the principle that such detection systems are feasible.

It is here that the benefits of embedded fiber sensors become very clear since if a signature-based approach is to be used, then the relative orientation of source and detector is critical. Preliminary experimentation with composite ultrasonic transducers and thin traveling wave piezoceramic sources also promise composite compatible excitation that can be permanently and economically embedded [7]. The prospect for a stored signature characterizing a particular structure and analyzed regularly by addressing the structure from a plug-in module is very real.

6.6 CIVIL PASSENGER AIRCRAFT—SOME QUALITATIVE OBSERVATIONS

Civil passenger aircraft are probably at the extreme of smart structures technology in every-day application. The actual sensing and actuation technologies and the structural design are relatively conservative. However, it is most certainly adaptive in that it is structurally different at take-off and landing than during the cruise, and it is definitely "smart" since it can change in responses depending on external circumstances.

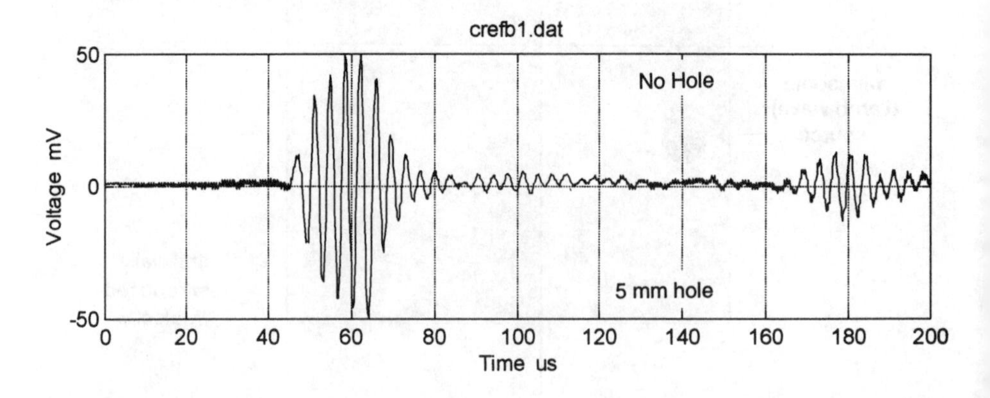

Figure 6.15 Typical trace from the apparatus shown in Figure 6.14. The hole is detected in the lower trace as the additional signal between the two major pulses.

Its adaptability is impressive. The aircraft automatically maintains a stability that is so good that a bump in turbulence equivalent to that experienced when hitting a medium-sized pothole in a car will make the airline passenger apprehensive. The aircraft can find its way from London to Auckland or Delhi to Bogota using its own navigational system and local guidance and tracking. It can take-off and, in many airports, can also land without any human intervention. In fact, the smoothest touch-downs are almost always those in which the automatic landing system takes control. It incorporates very comprehensive safety systems, optimizes its fuel consumption, and provides comprehensive climatic control. Every safety critical widget and gizmo exists in triplicate, and the whole thing is complemented by skilled staff.

The full technological story could be told but is actually well known within the engineering community. Civil transport aircraft combine conservative design with advanced technologies by insisting on triple redundancy, comprehensive maintenance schedules, and stringent procurement procedures—all in place to ensure that the needs of advanced technological solutions can be addressed with safety and confidence but at a very significant cost. The maintenance schedule on a jumbo jet runs to hundreds of thousands of dollars per annum. This then raises the question of whether, despite the aircraft's inherent "smartness," more can be gained by examining the procurement and maintenance procedures while retaining the all-important community confidence.

These factors are all well known, but it is interesting to speculate on the pressures that imposed expensive, active, and very comprehensive instrumentation solutions on the civil airline transport industry. My own view is that there are two principal driving forces: (1) Excess weight in the air is exceedingly expensive, so there is a continuing drive to operate the structure at virtually 100% full load and to optimize factors such as fuel economy; (2) Social costs of the civil airline industry are relatively easy to quantify and are very obvious, so civil air transport is the most legislated single aspect of our present infrastructure; furthermore, this legislation is internationally uniform.

The civil air transport industry is unique. This combination of the very high cost of carrying excess weight and the strictly imposed international legislative infrastructure ensures that costly and comprehensive intelligent structures approaches are desirable; indeed, necessary; and are finally very cost effective. The ever-expanding availability of low-cost and comprehensive instrumentation systems together with the ever-increasing awareness of social costs will, however, combine to drive the smart structures concept into other sectors.

6.7 SOME CONCLUDING COMMENTS

Structural instrumentation featuring arrays of sensors, complex data logging equipment, and debates concerning the Meaning Of It All have been with us for decades.

The aerospace industry has rigorously tested laboratory lash-ups of new concepts for half a century, and wind tunnel tests on ships and buildings have been conducted over a similar period. The aerospace industry has probably got it right, but the rest still have a lot to learn. A civil engineering colleague recently remarked that a fully loaded jumbo jet can continue in reliable service for thirty years while a bridge operating at less than 10% of its full load can fail dramatically. Despite my colleague's despair, the analogy is interesting and highlights radically different philosophies and, in particular, the approach of routine and legislated maintenance in aerospace compared to as needed maintenance in construction engineering.

Of course, one of the much vaunted potential benefits of the smart structures instrumentation in aerospace is that the routine maintenance can be replaced by as-needed attention. In civil engineering the concept is to apply the stitch in time by using the instrumentation for early warnings. Both approaches emphasize the necessity for the system to be reliable and predictable. In order to do this, the system must measure relevant parameters and feed these relevant parameters into both the design and maintenance process.

Complete integration of the instrumentation system into the design, build, and maintenance philosophy of the structure concerned is essential. Established design philosophies for all structural systems essentially assume that the structure is passive, will decay in accordance with past experience, and therefore is designed with some structural headroom. The uncertainties come in when new materials are used or the structure is applied in a new environment or is pushed into a new performance regime. Past experience is then replaced by conservative speculation but still retains its emphasis on considering only the structural materials.

If we bring in instrumentation as well and take it from the test bench to a continuous in-service assessment role, then the definition of the actual parameters that need measuring is critical. Structures decay through fatigue, erosion or corrosion, or occasionally through extremes of nature and rarely through anomalous loading since this last parameter is readily built into the original design. Perhaps then instrumentation systems should focus on these parameters despite the fact that in all the examples at the beginning of this chapter, the parameters that are measured are physical rather than chemical or environmental. In order to optimize the potential offered by an instrumented structure these factors need careful definition.

Technology also plays a critical issue. There is a rich and ever-expanding range of sensor types, and in particular the possibilities offered by distributed chemical sensing for corrosion conditions have only begun to be addressed. Sensors can also perform (and almost inevitably do) their own signal-processing and prefiltering functions, and the method by which to incorporate these into a system design process remains uncertain.

Rapidly changing technologies present both opportunities and problems. Certainly much more can be done—the aforementioned distributed chemical-monitoring systems are but one example. However, the need for assured reliability inhibits the penetration of new concepts into an extremely conservative intellectual environment. The counterargument is, of course, that perfection has not been achieved in the past and that somehow the potential benefits of new technologies must be evaluated against the potential risks of the unknown. The paradox has been largely resolved in aerospace by defining extensive qualification and procurement procedures, but in the more general civil structures environment equivalent procedures have been slow to emerge.

There remain some very real uncertainties. The aerospace industry, despite its procedures, is still struggling to define performance criteria for a neural network processor (many smart systems rely on such processors) that has effectively programmed itself against a training set that is difficult to control and oversee. Even if it were there is no guarantee that one neural network would program itself in exactly the same way as another, even given the same training set, since the role of noise in such networks is barely understood.

The word interdisciplinary is much abused. However, the whole smart structures arena is a genuinely multidisciplinary one that I have attempted to encapsulate in Figure 6.16. The questions start with the best value conflict since, invariably, the community desires far more than the manufacturer is prepared to manufacture at a price that the owner wishes to pay. Again much of this has been resolved in the aerospace industry, but in the general civil and transport sectors the trade-off of social cost against technical cost is very poorly understood. Meanwhile, at the technical level the interaction between the instrumentation system, the structural materials,

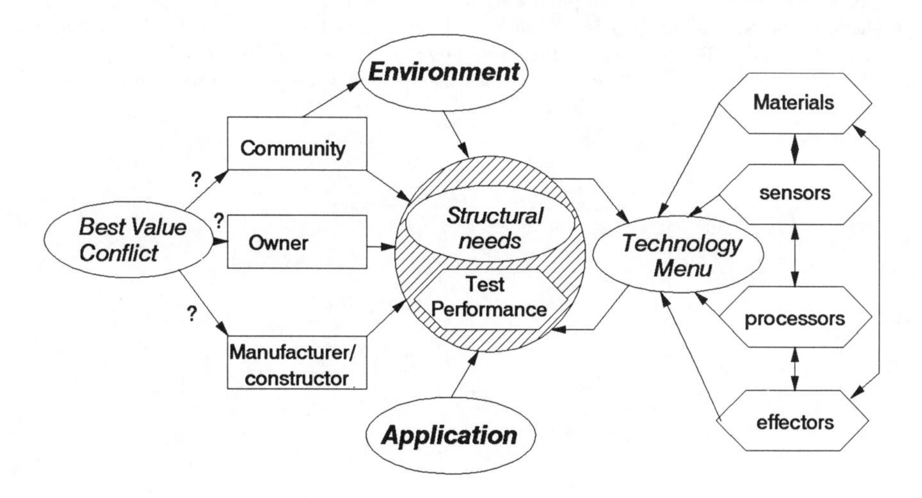

Figure 6.16 What is needed in the "smart structure"? An interdisciplinary problem!

and the structural properties continues to require careful characterization for individual cases. The greatest opportunities lie with the newest sensor systems for which the test procedures and acceptance criteria require definition.

The smart structure will have a great influence on structural design processes, maintenance procedures, and installation codes. The need for change is recognized within the broad community, and the necessary integration of technology, structures, and legislative and social environments is beginning. The best solutions need the best technology; the necessity for the best technology will ensure that, in time, this complete integration process will be implemented to maximum effect.

REFERENCES

[1] Milne, P. H., D. Carruthers, and A. McGown, "Integrated Systems of Deformation Measurement on Kingston Bridge, Glasgow," *Proc Engineering & Surveying 1994*, Institution of Civil Engineers, London.

[2] Sakamoto, M., and T. Kobori, "Control Effect During Actual Earthquake and Strong Winds of Active Structural Response Control of Buildings," *Fifth Int. Conference on Adaptive Structures*, Sendai, Japan, December 1994.

[3] Wada, B. K., "Basic Formulation for Adaptive Structures and Its Role in Space Structures," *Fifth Int. Conference on Adaptive Structures*, Sendai, Japan, December 1994.

[4] Geng, Z. J., G. G. Pan, W. S. Haynes, B. K. Wada, J. A. Gorba, "Six Degree of Freedom Active Vibration Isolation and Suppression Experiments," *Fifth Int. Conference on Adaptive Structures*, Sendai, Japan, December 1994.

[5] Culshaw, B., and P. T. Gardiner, "Smart Structures—the Relevance of Fibre Optics," *Fiber and Integrated Optics*, Vol. 2, 1993, pp. 353–373.

[6] Kim, J., and N. Stubbs, "Damage Localisation Accuracy as a Function of Model Uncertainty in the I-40 Bridge over the Rio Grande," *Proc SPIE 2446*, 1995, paper 2446–23.

[7] The EU Brite/Euram SISCO project in which the author's institution is a partner is investigating these concepts.

Chapter 7

Smart Materials

7.1 INTRODUCTION—ABUSE OF LANGUAGE

I wondered long and hard about including this chapter. I am certainly not a materials expert or a process chemist. Further I am very skeptical about the concept about the "smart" material, so it could be readily argued that this chapter is both presumptuous and superfluous.

However, I retain the hope that perhaps what follows will contribute a little to the debate on smart materials, will help to clarify the potential that is offered by materials in a responsive role, and may even assist in defining what we mean by such glamorous terms as "intelligence."

A useful way of viewing an intellectual hierarchy of materials and material structure should consider the following categories of materials:

- *Living materials* need to continuously dissipate energy and require suitable energy sources. They are capable of reacting to their surroundings, and these reactions may be classified into the *instinctive* and the *intelligent.* The latter implies a decision-making process based upon past experience. The former implies a reaction that requires no such analysis. The living material should also have the capacity to reproduce.
- The *instinctive or reactive material* is simply a transducer material that responds to a particular stimulus by changing its properties in a physical, chemical, electromagnetic, or similar domain. All materials are reactive especially to temperature.
- *Inert materials* are usually structural or decorative and are specifically designed to give minimal response to external stimuli.

These definitions can be debated at length but present some interesting features; for example, any manmade system that relies upon a computer in order to function is almost "living" apart from the question concerning its ability to reproduce, though possibly an automated PC factory meets the need entirely. The other two classes of materials do not require energy sources to function apart from an input stimulus. Another important observation is that no intelligent material has yet been realized from a single compound or a single structure whereas instinctive and inert materials are often pure compounds. Perhaps then, it is only appropriate to consider intelligent materials material systems rather than the materials alone.

An intelligent materials system must then have available to it, within the decision-making process, a data base and the necessary means of transmitting and processing the stimuli on which to base the decisions that the material system makes. It is interesting that possibly all intelligent material systems use some form or other of electronic charge transfer as the means for transmitting data. They also use electronic charge redistribution as the means of processing that data, though data storage and input stimuli can come from a variety of sources.

The purpose of this introductory discussion has been to highlight that intelligent materials can only exist as material systems while the instinctive and the inert may be single materials or material systems. In all cases, the material must fit the need as well as possible, so for all cases the real problems concern material selection and design.

7.2 MATERIALS AND MATERIAL SYSTEMS—WHAT THEY CAN DO

Materials and material systems can be rather arbitrarily characterized as inert, instinctive, and intelligent, reflecting more their application than their inherent properties (Figure 7.1). The inert material is meant to sit there and do its job. The instinctive provides a slavish input/output response, while the intelligent determines its output from a range of diverse inputs. The ability to respond to diverse inputs implies a commensurate diversity within the material structure, so intelligent materials, at least in this sense of the word, are always material systems.

Inert materials range from copper wires and optical fibers through permanent magnets to light structural alloys, composites, and decorative plastics. Their ability to respond to and/or process environmental parameters must, by virtue of their desired function, be minimized.

The instinctive material—often attributed with "intelligence"—includes all the transducer materials as discussed in Chapters 3 and 4 as well as a host of others, ranging from polymers that show their stress history through mechanically induced changes in polymeric structure through liquid crystals displays to photochromic sunglasses. There is also a host of instinctive material structures such as microsphere-encapsulated dyes or adhesives that release on impact. However, we are already into ambiguities since the simple fuse—most definitely an instinctive material—is not too far away from the inert conductor.

Intelligent material systems are always material systems rather than single substances and all require an energy source in order to process information from

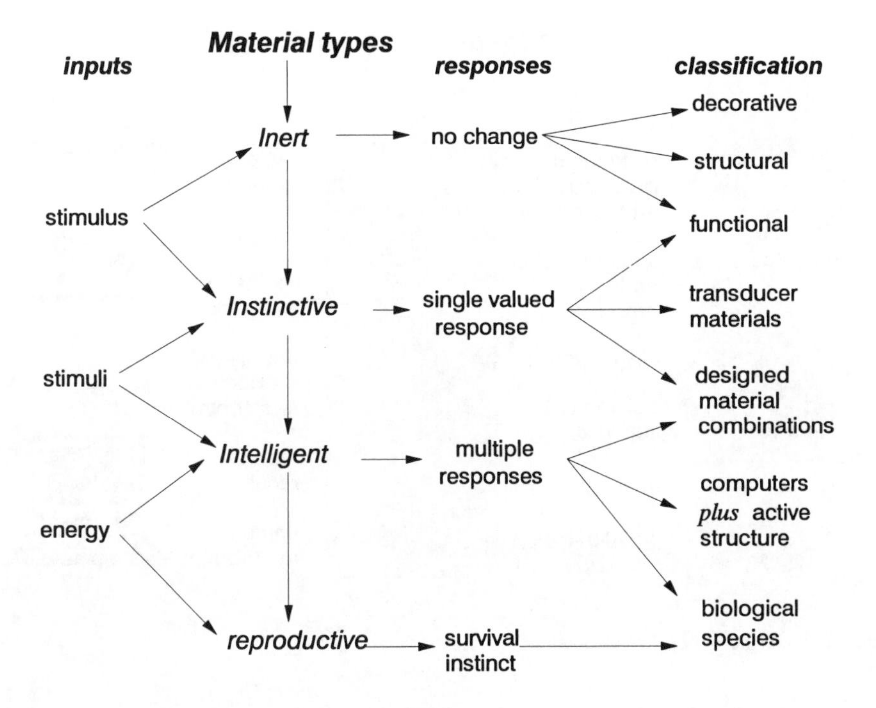

Figure 7.1 An idealized representation of material types.

their many sensory inputs. They also all require controlled energy release in order to execute the appropriate response.

Such material systems may be biological, and indeed the biological precursor is often cited as the ultimate aspiration for the materials system designer, though whether the engineer can or should attempt to replace evolution by a complete design philosophy can be endlessly debated.

The principal feature that the biological intelligent system enjoys but the computer-based equivalent has not yet developed is the capacity for growth, self-repair, and reproduction, but even the biological precursor has its limitations.

7.3 THE BIOLOGICAL ASPIRATION

There is more than a hint of Narcissus in viewing the biological model, invariably the mammal, and frequently the human, as the ultimate in material systems. I believe that our real aspiration is to incorporate the useful features of the human model with those already present in human invention (Figure 7.2):

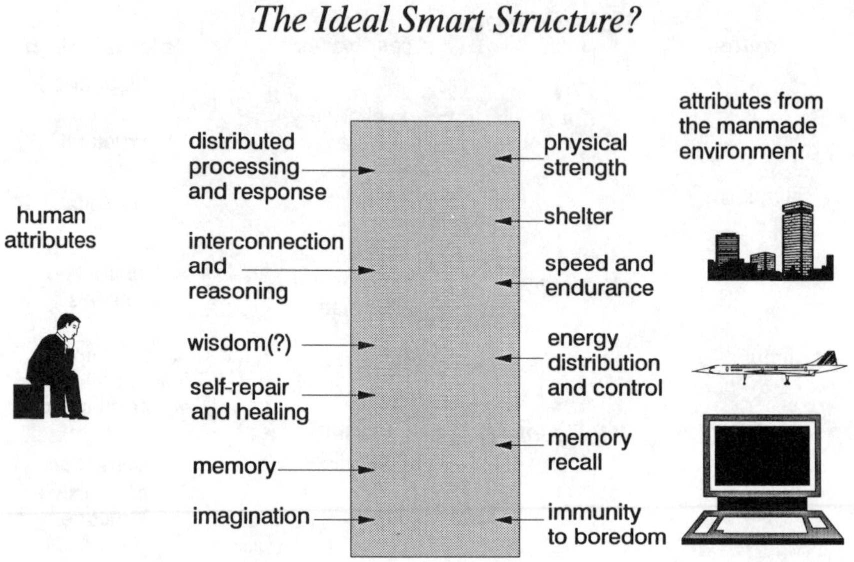

The Ideal Smart Structure?

human attributes

distributed processing and response
interconnection and reasoning
wisdom(?)
self-repair and healing
memory
imagination

attributes from the manmade environment

physical strength
shelter
speed and endurance
energy distribution and control
memory recall
immunity to boredom

Figure 7.2 Getting the best of all possible worlds?

- The human being enjoys vast memory storage but very poor memory recall. The human response has been to invent artificial memory, starting with tablets of stone evolving to the printing press and running now to the CD ROM.
- The human enjoys an immensely interconnected complex neural processor. This could, in principle, be emulated and, therefore could, in principle, reproduce human-reasoning capacity. All that restricts this is the available technology of three-dimensional complex interconnecting arrays.
- The human processor is incapable of extensive repetitive arithmetic processing. The abacus was invented to compensate this weakness, and this has evolved into the Cray computer.
- The human is environmentally fragile and, in particular, must accurately control body temperature.

Consequently, the human needs for shelter and transportation have necessitated the gradual evolution of a range of specific materials that often cannot be derived from biological sources but are processed from the earth's resources. These ensure that the human has available materials whose properties meet the need and, in particular, whose strength and hardness greatly exceeds that available in any biological substance.

Biological material systems are inherently complex and have been designed through evolution rather than CAD packages. While the biologist can tamper with this and short-circuit evolution by initially breeding and now genetic engineering, he still obtains material systems with the same generic properties but tuned slightly toward a particular need. The more traditional engineering disciplines are now beginning to aspire toward seeking these properties in synthetic systems and applying these to the social infrastructure of transport, shelter, and communications. Indeed, for the last of these, many could argue that the Internet has a life of its own, so perhaps we have already strayed into this blurred domain.

7.4 SOME CONCLUDING THOUGHTS

My view is that the long discourse on intelligent materials has really missed the point. Mankind's development has been marked by adapting the resources of his environment to provide functions that he is incapable of addressing himself but perceives as needs. These include the desire for stronger, lighter, and more durable materials. These materials must present no tangible hazards in use. The other principal driving need is for efficient means of transport and communication. Smart materials and structures must address these needs and do so by designing materials to suit a particular function and by incorporating within these materials structures the necessary intelligence to ensure their safe operation. In the intelligence context, man's deficiencies arise in memory retrieval and in man's ability to undertake repetitive and tedious tasks quickly and accurately, and these deficiencies are being rapidly redressed.

As for intelligent materials, these must surely address integrated detection, processing, control, and actuation functions. These already exist in the silicon semiconductor industry, though sometimes the actual detailed material processes themselves are not directly compatible. It does, however, seem that currently the limited scope for massive interconnect precludes the more advanced deductive reasoning capabilities that characterize even the simplest of birds or mammals. Perhaps polymer electronics will enable us to address these needs. We should recall, however, that specific material structures have been successfully designed to perform analogue computation within them, for example, for temperature compensation in pressure transducers. It is, therefore, relatively easy to argue that the smart material system does indeed already exist, though whether we shall ever meet the adaptive growth and reproduction capabilities of the famous biological precursor or whether this is either technically or ethically desirable remains to be seen. My belief is that intelligent materials are really about the intelligent design of useful material systems. Perhaps it is a rather touching human modesty that ascribes the intelligence to the material rather than to its inventor!

Chapter 8

Smart Structures and Materials—What of the Future?

8.1 DESIGNER ENGINEERING FOR FUTURE STRUCTURES

The principal objective here must be to evolve an integrated approach to the entire design process for engineering structures, incorporating not only structural engineering but also electronic and electromechanical systems targeted toward lifetime cost and reliability optimization. Large structures are complex entities and are limited by the availability of an all-embracing nerve system to probe every corner where corrosion, erosion, excessive loading, or material fatigue can take place.

Addressing these needs requires the realization of a number of diverse technical objectives and the integration of these objectives into an overall design procedure. The technical objectives are relatively straightforward to identify:

- Advanced structural materials engineering, including both passive design for purpose and adaptive design for crises: the current basic philosophy in structural engineering is to ensure the passive structure designs for crises.
- Sensor array techniques must be refined and developed to address the maximum structural complexity with the minimum sensor array. This entails both continued improvement and research into array techniques and technologies (for example, fiber optic distributed sensing and piezopolymer aerial integrating sensors) together with an engineering philosophy that optimizes their location with due recognition to structural parameters and signal-processing interpolation techniques.
- Array signal-processing techniques, incorporating artificial intelligence to obtain best guess interpretations of array signatures.
- Derivation of appropriate fuzzy structural models and experience-derived databases to assist in the data interpretation process.
- Actuator arrays to optimize structural response: these may be self-actuation, for example, tubes decaying or cracking and then releasing repair reagents, or

mechanical actuators or pumps to release appropriate chemical compounds at strategic points. These actuator array technologies are critical and, in particular, the distribution of energy to mechanical actuators with preferable conversion of this energy into mechanical actuation at the point where the actuation is needed, are all required. Current actuator technologies are only suited to actuation at a point.

- Distributed actuation will require new control algorithms and new models of system response especially since the actuation could indeed be mechanical, chemical, or even visual (for example, modifying electromagnetic radiation signatures). Optimizing this potentially multidimensional capability to produce the best structural response will require a fuzzy, nonanalytical approach broadly equivalent to the inverse of the signal-processing problem.

Figure 8.1 endeavors to encapsulate these processes and their interactions. The objective will continue to be to optimize structural performance with respect to consumer resources and user safety and comfort. The same basic philosophy will apply in transport, defense, civil infrastructure, and manufacturing. The current achievements will evolve into a totally revised approach to structural design. The

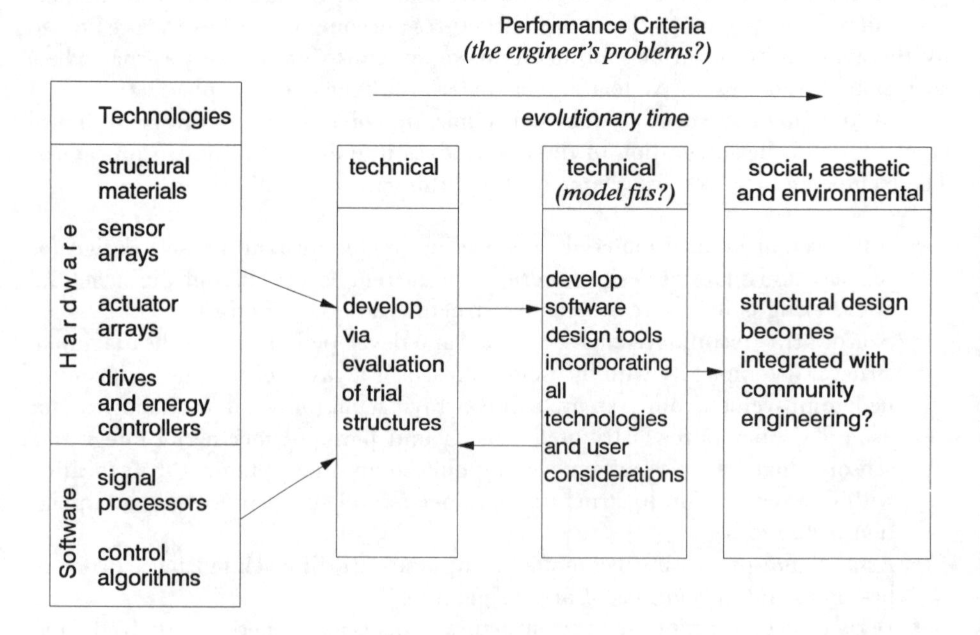

Figure 8.1 How structural engineering may evolve responding to the acceptance and exploitation of smart structural concepts and supporting materials technologies, hardware developments, and computational tools.

process is already beginning, but the necessary intellectual infrastructure is only now beginning to be recognized.

8.2 MATERIALS SYNTHESIS—THE INERT, THE INSTINCTIVE, AND THE INTELLIGENT

It is a somewhat glib observation, but everything is made of materials. Furthermore, everything is derived from natural materials and, even in our heavily industrialized world, natural materials are by far the greatest contributor to our overall environment. We attempt to control natural materials by pruning them, cultivating them, mowing them, and tinkering with their genetic origins.

We also go to considerable lengths to adapt and to reconfigure natural materials to our own ends, starting with stone-age man and continuing thereafter. In the context of intelligent materials (or should it be the intelligently designed materials?) our activities can probably be summarized as follows:

- The inert material should optimize the load-bearing capacity and/or aesthetic appeal of material structures. Most obviously this progresses through the use of composite materials but also incorporates aesthetic materials such as glasses and polymers and indeed novel treatments for natural materials especially wood and stone. This is the continuation of an evolution that goes back to the stone age.
- Instinctive materials that provide predictable and repeatable responses to given stimuli are a relatively recent addition to man's range of tools. They probably arose from the need initiated in the industrial revolution for control systems; and arguably the first instinctive material is the bimetallic strip that, while difficult to trace accurately, appears to have originated in the late nineteenth century. This simple observation indicates that there remains much to be learned about instinctive materials. Mankind does not assimilate information quickly. Despite the fact that the majority of technologists who ever lived are all alive today, it still takes generations for new ideas to really settle. We already see the optimization process for instinctive materials making steady progress with the general aim of enhancing the instinctive response to the appropriate stimulus and reducing it to any other stimulus. Therefore, we see composite transducers, temperature-tolerant magnetostrictives, and stress-sensitive polymers emerging as very important system building blocks. These instinctive materials will continue to develop in ever more complex forms with continuing improvements in understanding of material physics and commensurate development of refined processing technologies.
- Intelligent materials already exist, fulfilling the requirements of decision making and energy consumption in order to make these decisions from multidimensional

inputs. An intelligent material may be viewed as a collection of suitably interconnected and suitably weighted instinctive materials. Intelligent materials provide access to ever-increasing addressable memory systems and complex data bases and to rapid information-processing techniques. The limitations lie in the need for extensive interconnectivity to achieve a reasoning response. Intelligent material systems are even younger than instinctive ones. In the half century since the invention of the transistor, we have seen the unreliable and expensive devices of the 1950s evolve into inconceivably complex planar chip systems. There is no reason to assume that this progress should be stopped, though the biological interconnectivity means a three-dimensional rather than two-dimensional array and the technologists have barely begun to address interconnectivity in this light. When the three-dimensional processing problem is finally solved—the equivalent to the invention of the planar process in the middle of this century—even greater leaps in computing power will become almost immediately possible. The basic algorithms already exist.

The drive continues toward material structures whose responses begin to emulate the biological precursor from the perspective of furnishing a nervous system, processing signals from this nervous system, controlling distributed actuation and correction, and evolving the ability for self-repair. The objective continues to improve the performance of the man-made infrastructure and recognizes the vulnerability of inert materials in the physical infrastructure.

8.3 OF SOFTWARE, DESIGN, AND GENETICS

Computational tools will have a profound impact on future structural design. Our current design procedures are slow; the trial and error process essential for evolution still takes generations.

In the abstract, the processes of evolution and natural selection have been simulated by the artificial intelligence and artificial life communities by using genetic algorithms and evolutionary computation. Indeed, such simulations are in the most basic forms already available as computer games that emulate the rise and decline of cities or populations of creatures. Of course, the game succeeds because the boundary conditions imposed within the evolving system are essentially deterministic. However, these entirely synthetic populations do comprise the appropriate mixture of the intelligent and the instinctive and the inert material systems, so they have much in common with our aspirations in structural design.

The trend toward hybrid structures embracing passive structural elements, instinctive and reactive systems, and a modicum of intelligent control is already under way. Optimizing the combination of all these elements promises to be beyond the capacity of our current concepts of the design process based on human control of engineering rules. Inevitably, we shall more and more depend upon these newly developed computer tools to provide the design information. Perhaps there is a parallel here to our glib acceptance of computer memory and arithmetic capabilities that vastly extend beyond our own human potential. It is certainly not an outrageous step to suggest that the design process for complex systems can be similarly mechanized. That said, it is likely to take time and yet more computer power and algorithmic development. The computational fluid dynamics community deals with complex boundary conditions; and assure me, they need hours on a Cray to reach an answer that confirms experience. The overall boundary conditions for a fully smart structural design must be of comparable overall complexity. The CFD community does, however, have the advantage that their problem is relatively straightforward to formulate! The smart structures community (which one may argue embraces every aspect of our manmade physical infrastructure) is still seeking the basic rules!

Inevitably then, we must question the role of the structural system designer. Currently he interprets mechanical, technical, and environmental needs into a mechanical assemblage. When (and indeed it is when rather than if) the appropriate boundary conditions can be introduced with comparable confidence to those currently used in a computer game, the evolutionary process will proceed apace. These mechanical functions together with all the necessary passive, reactive, and intelligent components can be realized and optimized by machine. The structural system designer's responsibilities are now totally different. Certainly he or she must understand that the technicalities have been suitably specified, but then the technical responsibility stops and the total responsibility enters other domains. Aesthetics, social and community responsibilities, global resource management—indeed the whole social infrastructure—will then become dominant. Engineering and technical evolution will be compressed into a frighteningly short timescale. The inevitable conclusion is that profound changes within engineering and scientific disciplines are inevitable—and on a timescale measured in a very few generations.

8.4 APPLICATIONS AND CONFIDENCE

Those who historically design and build physical infrastructure—the civil, mechanical, and structural engineers—are inherently extremely conservative by nature. The physical infrastructure is vast and capital intensive and carries with it social and fiscal responsibilities to the community it serves. Consequently, there is little room for error, and overdesign is probably the norm. These inevitable responsibilities also enforce a "tried and true" education philosophy in future generations, and consequently the evolutionary process is rather slow.

This basic philosophy contrasts completely with that which has governed the evolution of the information age where rapid changes in techniques and standards are accepted as the norm. Indeed, the same is true of the smart materials fraternity, and both these enabling techniques and technologies must contribute to an emerging smart structures discipline. However, the "tried and true" design philosophy is beginning to be called into question since major infrastructure deterioration is causing a significant burden on society. This burden could be avoided given the availability of more information on the performance of the structure in question during its deterioration process. So the need for the smart structure is gradually being accepted. Confidence that this need can be addressed is also slowly developing given the first tentative steps that have been achieved via the case studies that were briefly described in Chapter 6.

I believe that the next decade will see the demonstration of very many more viable smart structure concepts, incorporating intelligent materials technologies into real, necessary, and desirable applications. This evolution will also influence design codes within the structural engineering professions and will also reconfigure educational philosophy within these areas, stimulating the need for generations of adaptable engineers capable of addressing complex design integration issues and of simultaneously stimulating the necessary technical progress in the appropriate disciplines.

The greatly encouraging feature within this observation is that many of these subdisciplines—or possibly all—are already at a sufficient state of maturity to provide a useful and convincing demonstration system that will serve to both develop user confidence and stimulate the necessary refinements in the contributing technologies. Perhaps then it will be possible for the conservative structural engineer and the speculative information and materials technologist to work together toward the common aim. There is much to gain for both sections of the community, and there is a sense of urgency imposed by the undeniable uncertainties in physical infrastructure. These combined motivations will ensure that smart structures and materials technologies will be major contributors to the engineering professions of the next century.

Bibliography

Here are just some of the references that may be useful for further reading:

Conference Proceedings

Annual North American Conferences on Smart Structures and Materials, Conferences on Fibre Optic Smart Structures and Skins, SPIE, Bellingham, Wash.

European Conference on Smart Structures and Materials, Glasgow 1992, 1994, Lyon 1996, SPIE, Bellingham, Wash.

International Conferences on Intelligent Materials, Lancaster, PA: Technomic Publishing.

International Conferences on Optical Fibre Sensors; most recently, *OFS,* Vol. 10, Glasgow 1994, Society of Photo-Optical Instrumentation Engineers.

Recent Advances in Adaptive and Sensory Materials and their Applications, Lancaster, PA: Technomic Publishing.

Journals

IEEE Press: many pertinent journals concerning topics such as computer applications, signal processing, nondestructive testing, instrumentation, and measurement.

Journal of Intelligent Materials, Structures and Systems, Lancaster, PA: Technomic Publishing.

Journal of Smart Materials and Structures, Institute of Physics, Bristol, United Kingdom.

Sensors and Actuators, Elsevier, New York.

General Texts

Cotterill, R., *The Cambridge Guide to the Material World,* Cambridge University Press, London, 1995.

Deutsch, S., and A. Deutsch, *Understanding the Nervous System,* IEEE Press, Piscataway, NJ, 1993.

Gandhi, M. V., and B. S. Thompson, *Smart Materials and Structures,* Chapman and Hall, London, 1992 (this book has an extensive bibliography on contributory technologies, especially SMAs, e-r fluids, and piezoelectrics, and a useful general bibliographic listing).

Lau, C., *Neural Networks,* IEEE Press, Piscataway, NJ, 1992.

Lynn, P., and W. Fuerst, *Digital Signal Processing with Computer Applications,* John Wiley, New York, 1989.

Michie, W. C., *Notes on Short Course on Smart Structures and Materials,* available from University of Strathclyde via the author.
Muller, R. S. et al., *Microsensors,* IEEE Press, Piscataway, NJ, 1991.
Rosen, C. Z. et al., *Piezoelectricity,* American Institute of Physics, New York, 1992.

About the Author

Brian Culshaw was born in Ormskirk, Lancashire, England, on September 24, 1945. He graduated with a BSc in physics in 1966 and a PhD in electrical engineering in 1970, both from University College, London. He joined Strathclyde University as professor of electronics in September 1983 after previous appointments as a postdoctoral fellow at Cornell University; a technical staff member at Bell Northern Research, Ottawa, Canada; a lecturer and, later, reader at University College, London; and a senior research associate in the Applied Physics Laboratory at Stanford University.

Culshaw worked on the design and technology of microwave and semiconductor devices until 1975, when his interests evolved into guided wave optics with particular applications in sensing, signal processing, and instrumentation. His interests include fiber-optic gyroscopes, hydrophones, accelerometers, temperature probes, strain and pressure measurement, sensors, and a host of other measurement systems, and he has also ventured into signal processing architectures and high-speed network design. It was from this background that his interests in smart structures evolved through the appreciation that guided wave optics could make a significant contribution to structural instrumentation. Subsequently, he has become involved in a number of projects on smart structures, especially in composite materials and civil engineering. Culshaw has written extensively on microwave semiconductors, fiber optics, and smart structures and materials, having authored or coauthored over 300 papers and five textbooks. He has also chaired major international conferences in these areas and currently acts as vice dean of the engineering faculty at Strathclyde and as a director of SPIE.

Index

The Artech House Optoelectronics Library

Brian Culshaw, Alan Rogers, and Henry Taylor, *Series Editors*

Acousto-Optic Signal Processing: Fundamentals and Applications, Pankaj Das

Amorphous and Microcrystalline Semiconductor Devices, Optoelectronic Devices, Jerzy Kanicki, editor

Bistabilities and Nonlinearities in Laser Diodes, Hitoshi Kawaguchi

Coherent Lightwave Communication Systems, Shiro Ryu

Electro-Optical Systems Performance Modeling, Gary Waldman and John Wootton

Elliptical Fiber Waveguides, R. B. Dyott

The Fiber-Optic Gyroscope, Hervé Lefèvre

Field Theory of Acousto-Optic Signal Processing Devices, Craig Scott

Frequency Stabilization of Semiconductor Laser Diodes, Tetsuhiko Ikegami, Shoichi Sudo, Yoshihisa Sakai

Fundamentals of Multiaccess Optical Fiber Networks, Denis J. G. Mestdagh

Germanate Glasses: Structure, Spectroscopy, and Properties, Alfred Margaryan and Michael A. Piliavin

High-Power Optically Activated Solid-State Switches, Arye Rosen and Fred Zutavern, editors

Highly Coherent Semiconductor Lasers, Motoichi Ohtsu

Iddq Testing for CMOS VLSI, Rochit Rajsuman

Integrated Optics: Design and Modeling, Reinhard März

Introduction to Electro-Optical Imaging and Tracking Systems, Khalil Seyrafi and S. A. Hovanessian

Introduction to Glass Integrated Optics, S. Iraj Najafi

Introduction to Radiometry and Photometry, William Ross McCluney

Introduction to Semiconductor Integrated Optics, Hans P. Zappe

Laser Communications in Space, Stephen G. Lambert and William L. Casey

Optical Control of Microwave Devices, Rainee N. Simons

Optical Document Security, Rudolf L. van Renesse

Optical Fiber Amplifiers: Design and System Applications, Anders Bjarklev

Optical Fiber Sensors, Volume I: Principles and Components, John Dakin and Brian Culshaw, editors

Optical Fiber Sensors, Volume II: Systems and Applicatons, John Dakin and Brian Culshaw, editors

Optical Interconnection: Foundations and Applications, Christopher Tocci and H. John Caulfield

Optical Network Theory, Yitzhak Weissman

Optical Transmission for the Subscriber Loop, Norio Kashima

Optoelectronic Techniques for Microwave and Millimeter-Wave Engineering, William M. Robertson

Reliability and Degradation of LEDs and Semiconductor Lasers, Mitsuo Fukuda

Semiconductor Raman Laser, Ken Suto and Jun-ichi Nishizawa

Semiconductors for Solar Cells, Hans Joachim Möller

Single-Mode Optical Fiber Measurements: Characterization and Sensing, Giovanni Cancellieri

Smart Structures and Materials, Brian Culshaw

Ultrafast Diode Lasers: Fundamentals and Applications, Peter Vasil'ev

For further information on these and other Artech House titles, contact:

Artech House
685 Canton Street
Norwood, MA 02062
617-769-9750
Fax: 617-769-6334
Telex: 951-659
email: artech@world.std.com

Artech House
Portland House, Stag Place
London SW1E 5XA England
+44 (0) 171-973-8077
Fax: +44 (0) 171-630-0166
Telex: 951-659
email: bookco@artech.demon.co.uk